Tech Talk

Better English through Reading in Science and Technology

FELIXA ESKEY

Ann Arbor
The University of Michigan Press

To my mother and father
for their inspiration

Copyright © by the University of Michigan 2005
All rights reserved
ISBN 0-472-03077-9

Published in the United States of America
The University of Michigan Press
Manufactured in the United States of America
♾ Printed on acid-free paper

2008   2007   2006   2005   4   3   2   1

Acknowledgments

I am deeply grateful to Professors Steve Molinsky and Marnie Reed of the Boston University School of Education for their encouragement of this project, their excellence in teaching, and their dedication to their students. I am also very grateful to Kelly Sippell of the University of Michigan Press for her patience and perseverance during the long permission and revision process, as well as for her invaluable suggestions. Many thanks to my husband for his constant support, his advice and help with all aspects of this undertaking, and his sense of humor. Finally, thanks to all my students for continuing to challenge me and for fueling my love of language teaching.

Grateful acknowledgment is made to the following individuals, authors, publishers, and journals for permission to reprint previously published materials.

Copyright Clearance Center for "Caught on Tape" by S. McDonagh, from *Science News*, June 7, 2003, copyright © 2003; "Getting a Grip" by K. Cobb, from *Science News*, August 31, 2002, copyright © 2002.

FotoSearch for photos of brain, bakery façade, and hot air balloon. Reprinted by permission.

Green Mountain Power for Wind Turbine image from *http://www.greenmountainpower.biz.*

Houghton Mifflin for material from *The Way Things Work* by David Macaulay. Compilation copyright © 1988 by Dorling Kindersley, Ltd. Text copyright © 1988 by David Macaulay and Neil Ardley. Illustrations copyright © 1988 by David Macaulay. Adapted and reprinted by permission of Houghton Mifflin Company. All rights reserved.

"National Academy of Engineering Reveals Top Engineering Impacts of the 20th Century: Electrification Cited as Most Important," press release, February 22, 2000, copyright © 2000. Reprinted by permission.

National Aeronautic and Space Administration (NASA) for photo of spacewalker. Reprinted by permission.

National Oceanic and Atmospheric Administration (NOAA) for photo of marine diatoms. Reprinted by permission.

National Public Radio, Inc., for material from Science Friday™, *www.sciencefriday. com:* "Smell," March 10, 2000; "Global Climate Change," February 18, 2000; "Engineering Feats," February 25, 2000; "Sounds Like Science: Fun Factoids," 1998–1999; "Sounds Like Science: Science Quiz," 2000. Science Friday™ © Samanna Productions, Inc.

National Space Biomedical Research Institute for text for "Nutritional Supplements May Combat Muscle Loss," copyright © August 27, 2002.

The New York Times for "Running Late? Researchers Blame Aging Brain" by Sandra Blakeslee, March 24, 1998, copyright © 1998; and graph "How Time Flies" by Dr. Peter Mangan, March 24, 1998, copyright © 1998.

New York Times Special Features for "Wind Power for Pennies" by Peter Fairley, from *Technology Review,* July/August 2002, copyright © 2002 Technology Review; "Ten Emerging Technologies that Will Change the World," from *Technology Review,* February 2003, copyright © 2003 Technology Review.

U.S. Department of Energy for photos of AGGDIS robot and wind turbines. Reprinted by permission.

U.S. Fish & Wildlife Service, Washington, DC Library, for photo of lizard. Reprinted by permission.

Contents

UNIT 7 Engineering Achievements

Introduction

Purpose of This Book

The intent of this book is to improve English language skills through realistic, interesting reading. This method strives to combine the benefits of pleasure reading and direct instruction by incorporating natural reading materials into the classroom. The book consists of authentic articles and excerpts, along with related exercises in comprehension, vocabulary, discussion, grammar, and writing.

The readings are meant primarily for those learners who are interested in science and technology, or those who are more motivated to read scientific and technical literature than other genres (such as fiction, biography, history, news, etc.). My own parents, who immigrated to the United States as middle-aged adults, are a good example. My mother enjoyed biographies and historical novels and began to read them in English, with increasing success. My father, on the other hand, was drawn to magazines about engineering and mechanics, which he read with focused interest. In addition, this book may be especially helpful to English learners who are engineers, scientists, or university students in technical fields. For them, such reading is not only a source of relevant vocabulary, but also a good example of technical writing style; and both vocabulary and style are provided within an interesting and familiar context. Of course, this book is by no means intended only for scientists and engineers. It is for everyone who wants to learn English and also happens to enjoy science. I also hope that this book will provide some additional motivation and encouragement to reluctant readers of other genres.

To those familiar with the research in second language reading, this volume does not constitute "free" reading in a strict sense since the materials are not actually chosen by the students themselves. Yet, the articles contained here are authentic texts, written in non-technical English, and are likely to be of interest to a wide audience. These selections exemplify the types of reading materials that are easily available to the public and common in academic study. Once students have been introduced to these types of materials, they will be able to select materials that are of particular interest to them and better satisfy academic requirements.

This textbook aims to:

▸ expose students to English language in a natural, meaningful context
▸ introduce students to interesting and relevant reading materials in science and technology

▸ increase vocabulary, both high-frequency and specialized, and demonstrate how it is used
▸ improve reading comprehension
▸ teach grammar through examples of real usage
▸ provide practice in the writing styles that are required in science and engineering—*explanation, persuasion, critical analysis*
▸ provide opportunity for verbal communication through class discussion
▸ encourage language learners to read independently and to learn language through reading

The material in this book is intended for high-intermediate to advanced students. Although the readings come from U.S. sources, they should be of interest to learners in other countries as well. This textbook is appropriate for all adults. It would be particularly useful to English students who are also scientists and engineers, including university students and professionals.

Materials

All materials are actual, unaltered English texts. They include articles from magazines and newspapers, excerpts from non-fiction books, and items from the Internet. Such materials are widely available and are easily accessible through public and university libraries. The readings span a variety of science-related topics, and many of them deal with the interaction of technology and society. They are meant to be enjoyable as well as to provide food for thought and discussion. The linguistic difficulty of the selections generally increases as the book progresses. Yet, the language is fairly general, rather than technically specialized, so as to provide the students with models and opportunities for everyday communication.

Reading

If class time allows, the articles should be read in class during a silent period, each student reading alone. This will simulate natural, free reading, allowing each student to make individual associations and interpretations. The length of the reading period will depend on the length of the selection. If class time is limited, students may complete their readings outside of class, after doing the pre-reading exercises in class.

Exercises

Each unit includes pre-reading and post-reading exercises. The pre-reading exercises set the stage for the reading by introducing the topic and key vocabulary, helpful to understanding the readings. These activities tap into what the students already know about the subject, put them at ease with it, and stimulate further interest in it. Post-reading exercises check reading comprehension, strengthen the acquisition of new vocabulary, and provide a forum for verbal discussion and written communication. The vocabulary taught and practiced are high-frequency words common to academic study. The discussion and writing activities deal with open-ended questions; they provide an opportunity for personally meaningful and creative use of language. At the end of each unit, the use of a certain grammatical form is pointed out and reviewed. Some units also offer suggestions for expansion activities that the students can do on their own outside of class; these provide additional exposure to language and reading materials.

Pre-reading exercises include:

▸ **Search Your Knowledge**—introduction to topic
▸ **Key Words**—vocabulary preview, if needed

Post-reading activities include:

▸ **What's the Point?**—reading comprehension
▸ **Understanding Words and Phrases**—vocabulary and idioms
▸ **Grammar Check**—brief illustration of a grammatical feature
▸ **Let's Talk about It**—verbal discussion
▸ **What Do You Think?**—written assignments
▸ **Activities for Fun**—independent activities outside of class

Class discussions, in pairs or larger groups, give students an opportunity to converse and to practice pronunciation and oral presentation in a meaningful context. For additional speaking practice, students may volunteer to present their written assignments. This may also stimulate further class discussion.

Answer Key

The answer key, available online for teachers (*www.press.umich.edu/esl/tm/*) provides answers to short-answer exercise questions. It also offers sample answers to some pre-reading exercises, where such answers help provide background knowledge or stimulate preliminary discussion.

How to Read

Here are some suggestions for you, the student, that will help you enjoy reading in English and increase your reading comprehension.

First, if applicable, think about how you read in your native language. Do you read silently or aloud? Do you know every word? What do you do when you encounter a word you don't know? If you don't understand every word, can you still understand what the text is about?

As you read each reading selection in this book, rely as much as you can on what you already know. There are several sources of this background knowledge: your familiarity with the general topic, the title of the article, illustrations, the language you already know, and your personal experience can help you figure out what the author is trying to say. You don't need to know every word to understand the main message of the text. Good readers often skip words they don't know, but they are still able to understand the meaning of a sentence or the rest of the text. You may be able to figure out the meanings of many new words and expressions from context.

In English writing, the title of a text and the first paragraph tell the reader what the text is about. Similarly, the first sentence of each paragraph reveals what the rest of the paragraph is about. Use this structure to make sense of what you read. Use your existing knowledge with new information in the text to comprehend an article or a story.

For a longer text, you may want to skim it first. To **skim** means to read or look at written material quickly in order to get the main idea. This will help you understand what you read. When you skim, take the following four steps.

First, read the title and subtitles, look at illustrations, and read illustration captions; this will help create a context for you.

Second, read the first one or two paragraphs to learn the general message of the article.

Third, read the last paragraph for the author's summary or conclusions.

Finally, read the first (and, if you wish, the last) sentence of each paragraph or section to get some supporting information for the main message. Remember that in skimming, the goal is to grasp the *main idea;* details are not important. After skimming, read the entire article.

Read quickly, but comfortably. Don't worry if you don't understand a word or a phrase right away—just keep reading. The text that follows will usually explain the meanings of new items. Even native speakers who are good readers don't understand every detail when they read. They guess meanings, and you should do the same. Words, phrases, grammatical forms, and ideas are often repeated in the text, and this helps readers become more certain about new meanings. When you have finished a section of a reading, you can, if you wish, re-read previous words and phrases to check your understanding and to become more familiar with new expressions.

Happy reading!

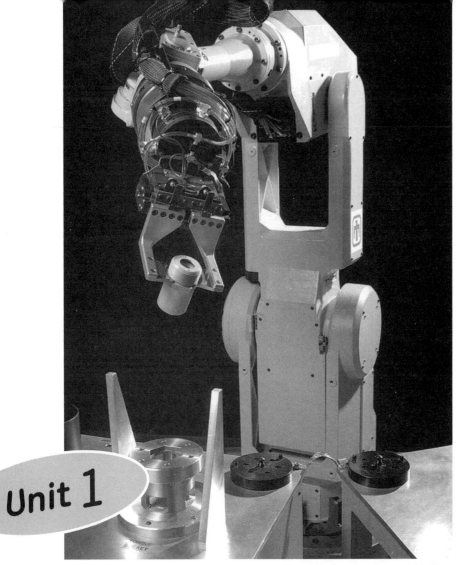

An industrial robot on an assembly line.

Computers and Automation

 Search Your Knowledge

1. How often do you use a computer? What do you use it for?

   List some examples on page 2 of how computers are used in everyday life.

Your Own Uses	Other Applications

2. How are computers like other machines? List some similarities.

3. How are computers different from other machines? List some differences.

Four short readings about computers follow. They are excerpted from the book *The Way Things Work* by David Macaulay. A comprehension check activity appears between the second and third readings.

Some of the words in the readings are underlined. These words may be new to you; therefore, they are underlined so that you can find them easily later on, if you wish to refer to them again. See if you can figure out what they mean from context or from the other words and meanings around the underlined word. These words are also included in the vocabulary exercises under Understanding Words and Phrases.

The boldfaced words in the texts are glossed in the margin. These non–high frequency vocabulary words or phrases are helpful to understanding the reading.

Machines with Memories

from *The Way Things Work*

Computers and calculators are a <u>revolutionary</u> development in the history of technology. They are fundamentally different from all other machines because they have a memory. This memory stores instructions and information.

In a calculator, the instructions are the various methods of arithmetic. These are <u>permanently</u> remembered by the machine and cannot be <u>altered</u> or added to. The information consists of the numbers <u>keyed in</u>.

A calculator requires an input unit to feed in numbers, a processing unit to make the calculation, and an output unit to <u>display</u> the result. A calculator also needs a memory unit to store the arithmetic instructions for the processing unit, and to hold the <u>temporary</u> results that occur during calculation.

Anatomy of a Calculator

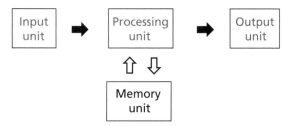

Program Power

A computer contains the same basic four elements as a calculator. It differs in that its memory can be given a different set of instructions, called a computer program, for different tasks. A program can turn a computer into, for example, a game player, a word processor, a paintbox, or a musical instrument. It instructs the processing unit how to perform the various tasks, and stores scores, words, pictures, or music.

Computer programs consist of long <u>sequences</u> of instructions that individually are very simple. The computer is instructed to distinguish two kinds of numbers and to put these two numbers into its memory. It is also told how to multiply, which it does by adding the first number to itself by the second number of times. Computers can perform millions of instructions in a matter of seconds.

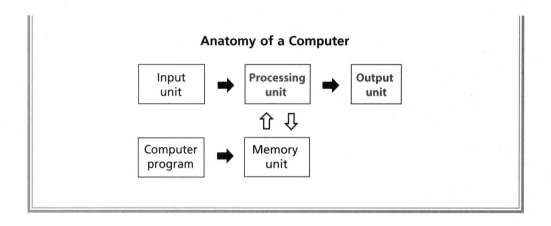

Anatomy of a Computer

What's the Point?

Show your understanding of the reading. Based on the text you just read, choose the best answer to complete these statements.

1. The most important difference between computers and other machines is that computers _____.

    a. are faster

    b. have a memory

    c. can print words

    d. can play a game

2. A computer needs a memory so that it can _____.

    a. display results

    b. understand numbers

    c. make calculations

    d. store instructions

3. A computer is different from a calculator because _____.

    a. a computer can be programmed to do different jobs

    b. a computer can do more difficult operations

    c. a computer can complete a task faster

    d. a computer can remember instructions

Understanding Words and Phrases

Determine the meaning of the underlined word in each sentence, and complete each related task. These sentences are not directly related to the text you just read. Use your own knowledge and experience—as well as what you learned from the text—to help you with this exercise.

The underlined words also appear in the text you just read; they are underlined in the text for easy reference. You can refer back to the text (page 3) to see how these words are used there. Using both contexts—the text and the sentences—can help you figure out what the words mean.

1. In the 16th century, Copernicus had the <u>revolutionary</u> idea that the earth orbited the sun.
   Explain why computers are a revolutionary development in technology.

2. When we first came to Boston, the hotel was our <u>temporary</u> home until we bought our own house; now we have a <u>permanent</u> home.
   In a computer, what kind of information is permanent and what kind is temporary?

3. His suit didn't fit properly, so he asked a tailor to <u>alter</u> it.
   Why would someone want to alter a computer program?

4. In a movie theater, a motion picture is <u>displayed</u> on a large screen.
   Give a few examples of "output units" that different computers use to display results.

5. People <u>key in</u> new information into a computer by pressing keys with numbers, letters, and other symbols, arranged on a keyboard (or a keypad on a calculator).
   What other instruments have keys and keyboards?

6. The months in a calendar are arranged from January to December in a <u>sequence</u>, one month following the other.

Give another example of something arranged in a sequence.

Binary Code

from *The Way Things Work*

A computer <u>appears</u> to work with things with which we are familiar—words composed of letters or pictures made of shapes in different colors, for example. But a computer turns all of these things into sequences of code numbers. These are not the decimal numbers we use, but binary numbers. Inside calculators and computers, code signals for decimal numbers, letters, screen positions, colors and so on flash **to and fro** between the various <u>components</u>. These signals are made up of numbers in binary code.

to and fro: back and forth

Binary Numbers

	8	4	2	1
0	0	0	0	0
1	0	0	0	1
2	0	0	1	0
3	0	0	1	1
4	0	1	0	0
5	0	1	0	1
6	0	1	1	0
7	0	1	1	1
8	1	0	0	0
9	1	0	0	1
10	1	0	1	0

Computers and calculators use binary code because it is the simplest number system. Its two digits—0 and 1—compare with our ten digits (0 to 9).

The table on the left side shows how binary numbers relate to decimal numbers. Reading from right to left, the ones and zeros in each column indicate whether or not the number contains 1, 2, 4, 8, and so on, doubling each time. 0101, for example, is

$$0 \times 8 + 1 \times 4 + 0 \times 2 + 1 \times 1 = 5.$$

Each binary digit (0 or 1) is called a bit.

Working with Addition

As computers and calculators work, they make calculations in binary arithmetic at lightning speed. Components called adders do the calculations, which all break down into sequences of additions. Subtraction, for example, is done by steps that involve adding and <u>inverting</u> numbers (changing each 1 to 0 and vice-versa).

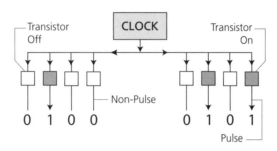

Code Signals

A binary code signal (*above*) is a <u>sequence</u> of electrical **pulses** traveling along wires. A device called a clock sends out regular pulses, and components such as transistors switch on and off to pass or block the pulses. One (1) represents a pulse, zero (0) a non-pulse.

pulse (physics): a brief burst of energy

Binary Arithmetic

There are only four basic rules:

A   0 + 0 = 0 and carry 0

B   0 + 1 = 1 and carry 0

C   1 + 0 = 1 and carry 0

D   1 + 1 = 0 and carry 1

	B	D	A	C
5	0	1	0	1
+ 4	0	1	0	0
9	1	0	0	1

1

Supermarket Checkout

from *The Way Things Work*

A supermarket checkout is a <u>sophisticated</u> input unit to the store's computer. Each product has a bar code that contains an identification number. The assistant moves the product over a window and an invisible **infra-red** laser beam scans the bar code, reading it at any angle. The number goes to the computer, which <u>instantly</u> shows the price of the product on the checkout display. The computer also adds up the check.

Checkout computers give the customer <u>rapid</u> accurate service, and they also <u>benefit</u> the store. The computer keeps a complete record of purchases and can work out **stock** levels. When the stock of a particular product falls, the computer can automatically order a new supply. The computer can also tell the store which products are selling well and which are less <u>popular</u>.

infra-red: electromagnetic radiation with a wavelength longer than visible light and shorter than microwaves

stock: items kept on hand for sale

Bar Codes

A bar code is a set of binary numbers. It consists of black bars and white spaces; a wide bar or space signifies 1 and a thin bar or space 0. The binary numbers <u>stand for</u> decimal numbers or letters.

There are several different kinds of bar codes. In each one, a number, letter or other character is formed by a certain number of bars and spaces. The bar code shown below uses five elements (three bars and two spaces) for numbers only.

START AND STOP CODES

These codes <u>precede</u> or follow the first and last number or letter in the whole code. By identifying these codes, the bar code reader is able to read a whole code either forward or backward.

FIVE-ELEMENT CODE

In this five-element bar code, the binary code 00110 <u>stands for</u> the decimal number 0.

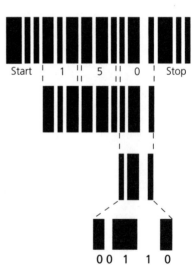

Laser Scanner

The checkout contains a laser scanner, which works in a similar way to a compact disc player. The laser fires a beam of infrared rays across the bar code, and only the white spaces in the code reflect the rays. The beam returns to the detector, which converts the on-off pulses of rays into an electric binary code signal that goes to the computer.

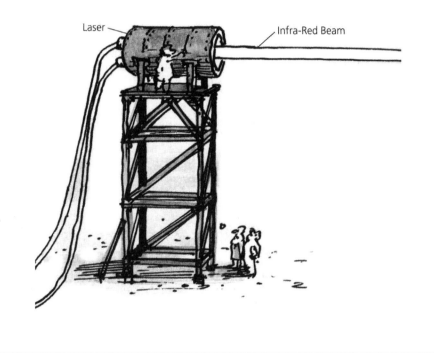

Laser   Infra-Red Beam

What's the Point?

Check your understanding of the texts you just read. Based on the reading, are the following statements true (T) or false (F)?

1. _____ Computers work directly with familiar words and pictures.

2. _____ The word *binary* means having two of something.

3. _____ Computers use binary code because they need a complex number system.

4. _____ The example bar code in this article represents the decimal number 5 by the binary code 10100.

5. _____ All calculations done by a computer are basically series of additions.

6. _____ In a computer, a component called the *clock* tracks the time of day.

 ## Understanding Words and Phrases

Based on the texts you just read, choose the best definition of the underlined word or phrase. These words and phrases are also underlined in the texts for your reference. If you aren't able to determine the meaning through context, use other vocabulary skills such as word parts and word forms to help you with the meaning.

1. to appear	a. to seem	b. to start	c. to try hard	d. to move
2. to invert	a. to add	b. to subtract	c. to turn around	d. to include
3. component	a. number	b. part	c. composition	d. computer
4. to precede	a. go after	b. go before	c. go together	d. go separately
5. to stand for	a. to represent	b. to contradict	c. to tolerate	d. to understand
6. sophisticated	a. interesting	b. intelligent	c. difficult	d. highly advanced
7. instantly	a. immediately	b. correctly	c. clearly	d. lightly
8. rapid	a. complete	b. unusual	c. very quick	d. polite
9. to benefit	a. be close to	b. give thanks to	c. give access to	d. be helpful to
10. popular	a. crowded	b. widely liked	c. personal	d. very cheap

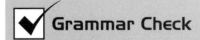 **Grammar Check**

The Passive

The first two readings in this unit contain the following sentences:

> "These [instructions] **are** permanently **remembered** by the machine and **cannot be altered** or **added to**."

> "The computer **is instructed** to distinguish two kinds of numbers. . . . It **is** also **told** how to multiply. . . ."

The words in bold type are verbs in the *passive* form. The form of the passive is **be + PAST PARTICIPLE** or **MODAL + be + PAST PARTICIPLE**. Notice that adverbs, such as *permanently* and *also*, appear between *be* and the **past participle**. How is the form of passive sentences different from the form of active sentences? When is the passive form used?

FORM: The object of an active sentence becomes the subject of a passive sentence. The subject of an active sentence follows the preposition *by* in the passive sentence; it is called the **agent** in the passive sentence. The two sentence forms have essentially the same meaning.

For example,

Active: The machine permanently *remembers* these instructions.

Passive: These instructions *are* permanently *remembered by* the machine.

Often, the *by*-phrase is not used in a passive sentence. This is because the exact agent is not known or is not important. These examples show the unnecessary agent in parentheses:

> These instructions *cannot be altered* or *added to* (by people).

> The computer *is instructed* (by someone) to distinguish two kinds of numbers.

> It *is* also *told* (by a programmer) how to multiply.

What are the active forms of these sentences?

Notice that passive sentences use only transitive verbs—verbs that can be followed by an object in an active sentence. It is impossible to use intransitive verbs (such as *sleep, fall, happen*) in passive sentences; this is because the active-form object, which becomes the passive-form subject, does not exist. For example, the following active sentence cannot be changed to a passive sentence: "She *sleeps* soundly at night."

Exercise 1

Change the active sentence to the passive. Use the by-*phrase only if it is necessary. Identify the sentences that cannot be changed to the passive because the verb is intransitive. Remember that in the form* **be** + **PAST PARTICIPLE,** *the verb be can have any tense—present, past, future, progressive, etc. In the form with modals, the past passive form is* **MODAL** + **have been** + **PAST PARTICIPLE.**

Example 1
Active: George **was preparing** the holiday dinner.
Passive: The holiday dinner **was being prepared** by George.

Example 2
Active: Someone **should have told** the students about the exam today.
Passive: The students **should have been told** about the exam today.

1. Some day, Roger's children **will run** the neighborhood hardware store.

2. All people **must obey** traffic laws.

3. Everyone **had eaten** breakfast before ten o'clock.

4. In two years, she **will become** a judge.

5. How **are** scientists **supposed to conduct** these experiments?

6. My sister **is going to pack and ship** the packages tomorrow.

7. We **don't understand** human behavior very well.

8. Workers **should have completed** construction of this building last month.

9. People **don't** usually **ski** in the summer.

10. **Has** a doctor **examined** the patient yet?

11. When **is** Lara **supposed to go** on vacation?

12. **Could** a single person **have lifted** this stone?

USAGE AND MEANING: The passive is used in the following situations.

1. The specific agent is not known or is not important. In this case, the passive form is used without the *by*-phrase. This is the most common use of the passive. It is especially common in technical and scientific writing. In such formal writing, the most important thing is to describe events and results. It is not important to know

exactly who did those things. It is understood that the agents are people—scientists, engineers, researchers, doctors. For example,

> Their children **were educated** in Italy.
>
> This university **was established** in 1861.
>
> Antibiotics **are used** to treat infections.
>
> A five-year research study **is being conducted** to evaluate the new test procedure.

2. It is important to identify the agent, or to distinguish a particular agent from any other. In this situation, the *by*-phrase is used, and it places the agent at the end of the sentence, a place of emphasis. For example,

> The special physics lecture **will be presented** by Professor McCune.
>
> These houses for the poor **are built** by volunteers. [*volunteers,* not the government]

3. Sometimes, the speaker uses the passive with the *by*-phrase in order to focus attention on the subject—or topic—of the sentence. (The sentence subject is usually the topic of the larger discussion, identified in the previous context.) This also serves the purpose of placing new information about the old, familiar topic at the end of the sentence, a place of stronger emphasis. Thus, a clear sentence begins with a familiar topic and ends with new information, which is recognized more easily at the end. Consider these examples from the last two readings in this unit. The first sentence creates the context, while the next sentence is passive:

a. "Components called adders do the calculations, which all break down into sequences of additions. Subtraction, for example, *is done* by steps that involve adding and inverting numbers."

> *Familiar topic:*   "subtraction," a type of calculation
>
> *New information:*   "adding and inverting numbers"

b. "There are several different kinds of bar codes. In each one, a number, letter or other character *is formed* by a certain number of bars and spaces."

Familiar topic:	bar codes representing "numbers, letters, or characters"
New information:	"a certain number of bars and spaces"

What is the corresponding active form of the second sentence in examples *a* and *b* above? Does the active sentence sound equally clear and logical to you in the given context?

Exercise 2

Now, change each passive sentence to a possible active form. Which form sounds better, more natural, to you? Why do you think the passive form was chosen for these items?

1. The city library was destroyed in the war. It was rebuilt ten years later.

2. First, the chicken is thoroughly washed. Then, it is seasoned with spices and placed in the oven to roast.

3. The holiday show was a great success and involved everyone. The singing was performed by the girls, while the tap-dancing was done by the boys.

4. The bread and pastries in this resort hotel are baked fresh every morning.

5. I have two kinds of green tea. One was given to me by a friend from Japan; the other was brought by a friend from China.

6. When were you first introduced to classical music?

7. Usually, this type of surgery is performed by a specialist, but this time, it was done by my family doctor.

8. Electricity is frequently produced from fossil fuels. It can also be produced from nuclear and solar energy.

9. Vulcanized rubber, which is used for automobile tires, was invented by Mr. Charles Goodyear.

10. This museum contains works of art from all over the world. The paintings were done mostly by European and American artists. The sculptures and pottery were created by Asian and African artists. The special collection of textiles was donated by Native American tribes.

Exercise 3

Make it personal! Write your own sentences describing your native country, its popular traditions, the school system, or your family customs. Decide which form—active or passive—is the best for a particular sentence.

1. _____

2. _____

3. _____

4. _____

5. _____

6. _____

Let's Talk about It

1. How does an automated bar-code checkout help the customer? Give two examples from the text. Can you think of other advantages to the customer?

2. How does a bar-code checkout help the supermarket? Give at least two examples from the text, and add your own examples.

3. Are computers helpful or harmful to our society? Why?

What Do You Think?

Do you think that computers and automation have improved your life? Write a few paragraphs arguing how computers have made your life better or worse.

A hot air balloon lifts off.

Flight

Search Your Knowledge

1. What are some ways in which people are able to fly?

2. Have you ever been in a small airplane? A glider? A helicopter? A balloon? If you have, how did it feel? If you haven't, how do you imagine it would feel?

3. List some reasons why it is difficult for human beings to fly.

18

4. What makes a balloon stay in the air?

5. How is an airplane different from a balloon? What makes an airplane fly?

Next, you will read about some principles that are important in flying and about different kinds of flying machines. The readings are excerpts from the book *The Way Things Work* by David Macaulay.

Some of the words in the readings are underlined. These words may be new to you; therefore, they are underlined so that you can find them easily later on, if you wish to refer to them again. See if you can figure out what they mean from context or from the other words or meanings around the underlined word. These words are also included in the vocabulary exercises under Key Words and Understanding Words and Phrases.

The boldfaced words in the text are glossed in the margin. These non–high frequency vocabulary words or phrases are helpful to understanding the reading.

KEY WORDS

What is the meaning of the underlined word in this example?

The men had to use a lot of <u>force</u> to move this heavy piano.

What is *force?* Can you name some forces in physics or in other areas? Make a list of them.

Heavier-than-Air Flight

from *The Way Things Work*

A kite, a glider, and a powered aircraft are three quite different ways by which an object that is heavier than air can be made to fly.

Like balloons and airships, heavier-than-air machines achieve flight by <u>generating</u> a <u>force</u> that <u>overcomes</u> their weight and which supports them in the air. But because they cannot <u>float</u> in the air, they work in different ways from balloons.

Kites <u>employ</u> the power of the wind to keep them **aloft,** while all winged aircraft, including gliders and helicopters, make use of the airfoil and its power of lift. Vertical **take-off** aircraft direct the power of their jet engines downward and heave themselves off the ground by **brute force.**

The two principles that <u>govern</u> heavier-than-air flight are the same as those that propel powered vessels—action and reaction, and suction. When applied to flight, suction is known as lift.

aloft: high up

take-off: the act of rising in flight (verb: *to take off*)

brute force: a force or effect that is purely physical and strong

KITE

A kite flies only in a wind, and it is held by its string so that it deflects the wind downward. The wind provides the force for flight. It exerts a reaction force that equals the pull of the string and supports the kite in the air.

Reaction Force

Pull of String

Wind

AIRFOIL

The cross-section of a wing has a shape called an airfoil. As the wing moves through the air, the air divides to pass around the wing. The airfoil is curved so that air passing above the wing moves faster than air passing beneath.

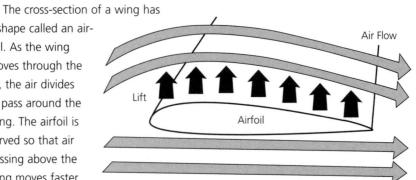

Fast-moving air has a lower pressure than slow-moving air. The pressure of the air is therefore greater beneath the wing than above it. This difference in air pressure forces the wing upward. The force is called lift.

GLIDER

A glider is the simplest kind of winged aircraft. It is first pulled along the ground until it is moving fast enough for the lift generated by the wings to exceed its weight. The glider then rises into the air and flies. After release, the glider continues to move forward as it drops slowly, pulled by a thrust force due to gravity. Friction with the air produces a force called drag that acts to hold the glider back. These two pairs of opposing forces—lift and weight, thrust and drag—act on all aircraft.

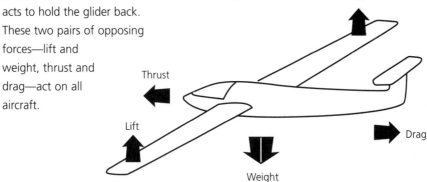

Flying Machines

from *The Way Things Work*

Many different flying machines now fill our skies. They range from <u>solo</u> sports and aerobatic planes to wide-bodied and supersonic jet airliners which carry hundreds of passengers. Some, such as pedal-powered planes, lumber along just above the ground, while others, such as **reconnaissance** aircraft, streak at three times the speed of sound at a height three times that of Mount Everest.

There are also unpowered gliders, of which the returning space shuttles are the largest and hang gliders the simplest. Development in other directions has led to helicopters and vertical take-off aircraft which are capable of rising vertically and **hovering** in the air. There are also kites of all shapes and sizes, some large enough to carry a person.

Machines also fly through water. Hydrofoils flying through the waves <u>employ</u> exactly the same principles that keep winged airplanes aloft.

reconnaissance: exploration of an area, especially to obtain military information

hovering: staying floating in the air over a particular place

GLIDER

Being unpowered, a glider cannot travel fast and so has long straight wings that produce high lift at very low speed.

SPACE SHUTTLE

The space shuttle re-enters the atmosphere at very high speed, and so has a delta wing like a supersonic airliner. It then glides to a high-speed landing.

HANG GLIDER

The A-shaped wing <u>inflates</u> in flight to produce an airfoil with low lift and drag, giving low-speed flight with a light load.

LIGHT AIRCRAFT

Short straight wings produce good lift and low drag at medium speed. Propellers or jet engines provide the power that produces the lift.

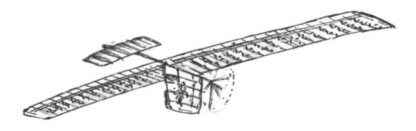

PEDAL-POWERED PLANE

Because the flying speed is very low, long and broad wings are needed to give maximum lift. Drag is at a minimum at such low speeds.

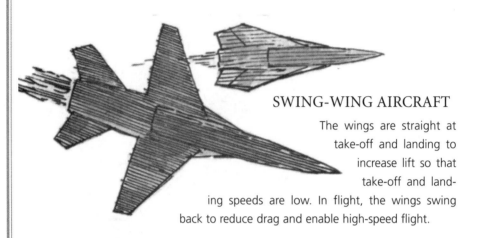

FORWARD-<u>SWEPT</u> WINGS

This experimental design gives high lift and low drag to produce good maneuverability at high speed. Two small forward wings called canards aid control.

SWING-WING AIRCRAFT

The wings are straight at take-off and landing to increase lift so that take-off and landing speeds are low. In flight, the wings swing back to reduce drag and enable high-speed flight.

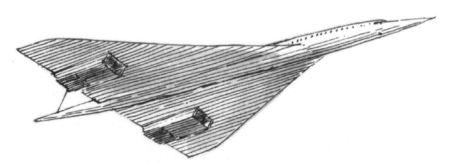

SUPERSONIC AIRLINER

Aircraft that fly faster than the speed of sound often have dart-shaped delta wings. This is because a shock wave forms in the air around the aircraft, and the wings stay inside the shock wave so that control of the aircraft is <u>retained</u> at supersonic speed. Take-off and landing speeds are very high as lift is low.

AIRLINER

<u>Swept</u>-back wings are needed to minimize drag at high speed. However, lift is also reduced, requiring high take-off and landing speeds.

FLAPPING WINGS

This is a highly efficient wing design that you should **look out for,** particularly in places where bird feeding is encouraged.

look out for (phrasal verb): watch carefully for something, especially something dangerous; be careful about (something)

What's the Point?

I. Show your understanding of the reading. Based on the text, pictures, and captions in this unit, give short answers to the following questions.

1. What objects can stay in the air because they weigh less than air?

2. What three types of flying objects are heavier than air?

3. How do all flying objects achieve flight, in general?

4. What makes kites stay in the air?

5. What makes gliders, airplanes, and helicopters stay in the air?

6. What methods do powered aircraft use to produce power that creates lift? Name three different methods.

II. Check your understanding of the text. Based on the reading, are the following statements true (T) or false (F)?

1. _____ Winged aircraft can fly because pressure difference produces suction on the wing.

2. _____ Kites can fly even when there is no wind.

3. _____ Aircraft wings are able to generate lift because of their geometric shape.

4. _____ Gliders have their own engines to produce power.

5. ____ Airplanes cannot fly faster than twice the speed of sound.

6. ____ The space shuttle flies like a supersonic glider.

7. ____ Wings that point forward allow easy movement in different directions at high speed.

8. ____ Wings that are swept back reduce drag at high speed and increase lift.

9. ____ Long wings help increase lift.

10. ____ Delta wings on supersonic aircraft prevent the formation of a shock wave.

💡 Understanding Words and Phrases

I. Choose one of the words to fill in the blank in each of the following sentences. You may have to change the form of the word to fit the sentence.

exceed	generate	retain
float	overcome	solo

1. Many power plants in the United States _____ electricity by burning coal or oil.

2. Boats, ships, and wood _____ on a river because they weigh less than water, but rocks sink to the bottom.

3. Patience and hard work help a person _____ difficulties in life.

4. Sylvia has always performed with her group, but tonight she will sing _____, and everyone's attention will be on her alone.

5. The price of the car _____ the amount of money I could spend, so I couldn't buy it.

6. Margaret decided to update her wardrobe, so she got rid of all her old clothes and _____ only her favorite hand-knit scarf.

II. Many English words have multiple meanings. These meanings are often related. However, only one of the meanings is usually correct in a particular situation or context. These sentences contain underlined words that you have seen in the texts. Each sentence is followed by two correct definitions of the underlined word. Choose the one definition that is appropriate for the given sentence.

1. Kites <u>employ</u> the power of wind to keep them aloft.

   a. to have a person working for payment

   b. to use

2. You can <u>inflate</u> party balloons by blowing air into them.

   a. to fill and swell with a gas

   b. to raise or increase too much; to exaggerate

3. Laws of physics <u>govern</u> flight.

   a. to administer public affairs or to rule a society

   b. to regulate or control

4. Some small birds can <u>sweep</u> their wings forward to achieve faster movement in the air. (Note: The past tense of *sweep* is *swept.*)

   a. to clean or clear away (dirt or snow, for example) with a broom or brush

   b. to move with a flowing motion

5. <u>Friction</u> between a glider and the air produces drag, which holds the glider back.

   a. the rubbing of one surface against another

   b. a conflict between two people because of disagreement

For each underlined word, write your own sentence using the other meaning, the one that you did <u>not</u> choose above.

1. employ: _____

2. inflate: _____

3. govern: _____

4. sweep: _____

5. friction: _____

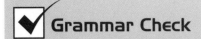 Grammar Check

Adverb Clauses

The excerpts you just read contain these grammatical structures:

a. "**Because** *they cannot float in air,* they work in different ways from balloons."

b. "**As** *the wing moves through the air,* the air divides to pass around the wing."

c. "It is first pulled along the ground **until** *it is moving fast enough for the lift generated by the wings to exceed its weight.*"

d. "Kites employ the power of the wind to keep them aloft, **while** *all winged aircraft, including gliders and helicopters, make use of the airfoil and its power of lift.*"

The part of each sentence that is in italics is an **adverb clause,** a kind of dependent clause. A *clause* is a group of words that has a subject and a verb. A *dependent clause* cannot stand alone as a sentence; it must be attached to an *independent, or main, clause.* In *a,* the group of words "they work in different ways from balloons" is the main clause and can stand alone as a complete sentence; it contains the main subject and main verb of a sentence. The group of words "because they cannot float in air" is a dependent clause. It is called an **adverb clause** because it functions like an adverb: It gives additional information about the main clause, particularly the main verb.

In these examples, the words *because, as, until, while* introduce an adverb clause. These special words are called **subordinating conjunctions,** because they connect a dependent, or subordinate, clause to an independent clause.

Notice the punctuation in these examples. When an adverb clause comes before the main clause, as in examples *a* and *b,* a comma is used to separate the two clauses. When an adverb clause comes after the main clause, as in example *c,* usually there is no comma. However, with the conjunctions *while* and *whereas,* as in example *d,* a comma is usually used even when the adverb clause follows the main clause. Also, when the conjunction *only if* comes in the beginning of a sentence, the subject and verb in the main clause are inverted, and a comma is not used. For example:

> **Only if** the animal is threatened *does it* attack people.

Here are some common subordinating conjunctions and their general meanings.

Time:	*as, until, before, after, since, when, while, by the time, as soon as, once, as long as, whenever*
Cause and effect:	*because, now that, since*
Contrast:	*although, though, even though, while, whereas* (more formal than *while*)
Condition:	*if, only if, even if, in case, in the event that, whether or not, as long as, unless*

Exercise 1

Find the adverb clauses in the following items. Underline each adverb clause, and circle the conjunction that introduces it. Punctuate the sentences with commas, periods, and capital letters, as needed. Watch out— some items contain more than one independent clause (sentence), and some contain more than one adverb clause!

1. By the time you come home from work dinner will certainly be ready.

2. We're enjoying our sightseeing vacation it hasn't rained at all since we got here.

3. Larry always carries an umbrella in case it rains.

4. Although I don't like vegetables very much I eat them every day because they're good for me.

5. When she is thirsty she prefers hot drinks whether or not it is cold outside.

6. Now that Mark and Christina have children they seldom go out to dinner or the movies.

7. Ken and Belinda have different tastes in music he likes country whereas she prefers jazz.

8. Since we are making good progress on this project let's keep working we can have lunch later unless you are very hungry now.

Exercise 2

Each item contains two independent clauses. Combine them into one sentence using the given conjunction in parentheses. Use commas where needed.

1. (**as soon as**) He gets home. He takes off his shoes.

2. (**because**) Tracy is very upset. She lost the diamond ring that belonged to her grandmother.

3. (**before**) You should brush your teeth. You go to bed.

4. (**only if**) Billy can have ice cream for dessert. He eats all his carrots.

5. (**since**) You insist on helping me with housework. You can vacuum the rugs.

6. (**as**) Henry admired the architecture around him. He walked along the street.

7. (**even though**) Mr. Wright is 84 years old. He still enjoys downhill skiing.

8. (**in the event that**) You cannot reach me at work during the day. Please call me at home.

Exercise 3

Complete the sentences with your own words. The subordinating conjunctions are in boldface. Be sure to add commas where needed!

1. Molly goes for a walk every day **even if** _____

_____.

2. Jerry doesn't like gardening **though** _____

_____.

3. **Once** Mrs. Bradley learned enough French _____

_____.

4. Her health will continue to improve **as long as** _____

_____.

5. **After** I finish this exercise _____

_____.

6. He will definitely buy this house **unless** _____

_____.

7. **Whenever** I feel sad _____

_____.

8. New York is _____ **while** _____

_____.

Let's Talk about It

1. What is the fastest flying machine included in the text? Why do you think this is true?

2. What is the slowest machine? Why?

3. Of the machines presented, list the ones that you think are used for public or personal uses. Why do you think so?

4. List the aircraft that you think are used for military purposes. Why do you think so?

5. What kinds of wings do birds have?

What Do You Think?

In your opinion, has supersonic flight changed the way people live and work? Write a short essay (two to three paragraphs) supporting your opinion.

On June 3, 1965, Edward H. White II became the first American to step outside the Gemini spacecraft, as shown here. More recently, on August 3, 2005, Stephen Robinson was tethered to a robotic arm in space.

Health Care in Space

Search Your Knowledge

1. Why might it be dangerous to send human beings into outer space? What are some problems with living in space that we do not have on earth?

2. Can you think of some health problems related to being in space for a long time? What could go wrong?

3. What happens when a person isn't able to exercise or move for a long time? What happens to the muscles and bones?

4. How can scientists learn about these problems and their possible solutions?

The text that follows Key Words was released by the National Space Biomedical Research Institute (NSBRI). This organization conducts research on health risks related to long-duration missions in space. Can you understand the underlined words in the text? These words may be new to you; therefore, they are underlined so that you can find them easily later on, if you wish to refer to them again. See if you can figure out what they mean from context or from the other words and meanings around the underlined word. The words are also included in the vocabulary exercises under Key Words and Understanding Words and Phrases.

The boldfaced words in the text are glossed in the margin. These non–high frequency vocabulary words or phrases are helpful to understanding the reading.

KEY WORDS

These sentences contain underlined words, which also appear in the article you are about to read. Knowing these words will help you understand the article. Can you figure out what these words mean from the sentences? In addition, answer the question or questions that follow each sentence.

1. Most milk sold in the United States contains two nutritional supplements: vitamins A and D.
   What other nutritional supplements do you know?

2. After the accident, Mr. Brown spent the rest of his life in a wheelchair; eventually, the muscles in his legs atrophied, becoming smaller and weaker.
   What parts of the body can atrophy from the disease diabetes?
   What organs can atrophy from drinking too much alcohol?

3. Laboratory mice are often the <u>subjects</u> of scientific and medical experiments.

   Have you ever been a subject in a research experiment? If so, describe it.

4. <u>Protein</u> is an important part of your diet because it is necessary for building muscles.

   Name some foods that are good sources of protein.

5. The <u>hormone</u> insulin is produced by your body and controls the level of sugar in your blood.

   What other hormones are you familiar with?

6. Here is an explanation of a technical term that may be useful to you:

   An *isotope* is a variation of a chemical element. Isotopes of an element have the same number of protons inside the atom, but they have a different number of neutrons. For example, Carbon-12 and Carbon-14 are isotopes of the element carbon; they both have six protons, but carbon-12 has six neutrons and is a *stable isotope* (not radioactive), while carbon-14 has eight neutrons and is a *radioactive isotope*. (Carbon-14 is used in "carbon dating" to figure out the age of objects, such as archeological finds.)

NEWS RELEASE

Nutritional Supplements May Combat Muscle Loss

from the *National Space Biomedical Research Institute*

HOUSTON—(Aug. 27, 2002)—Early <u>indications</u> show that nutritional <u>supplements</u> may lessen muscle <u>atrophy</u> brought on by space travel, prolonged bed <u>confinement</u> or immobility.

To study space travel's effect on muscles, Dr. Robert Wolfe of the University of Texas Medical Branch at Galveston enlisted healthy <u>subjects</u> to stay in bed 28 days during a National Space Biomedical Research Institute study.

"One cause of muscle atrophy in space is lack of muscular activity. That's why bed rest is a good model because it minimizes activity, and like astronauts, you lose muscle mass primarily in the legs," said co-investigator Dr. Arny Ferrando, a professor of surgery at UTMB and Shriners Hospital for Children in Galveston. "When muscles are inactive, as they are in space, they don't make new <u>proteins</u>. If muscle <u>breakdown</u> rates are the same, that means you lose muscle."

Researchers are <u>attempting</u> to increase protein <u>synthesis</u> rates with <u>supplements</u> of amino acids, which are the <u>raw</u> materials of <u>protein</u>. Participants received the <u>supplements</u> three times a day, and researchers compared the protein <u>synthesis/breakdown</u> rates and muscle mass before and after the bed-rest study. This data was compared to results from a control group that received a placebo drink instead of the <u>supplements</u>.

"Early results suggest that the amino acid <u>supplement</u> is able to maintain <u>synthesis</u> rates and body mass," Ferrando said.

During the study, <u>subjects</u> must remain in bed and can get up only briefly to use a bedside commode. They eat and bathe from their beds, and daily activities <u>encompass</u> watching television, reading books and using a bedside computer.

Midway through the study, researchers <u>determine</u> muscle mass and function by testing the <u>subjects</u>' strength and body composition.

They gather the most vital data, the <u>protein</u> <u>synthesis</u> and <u>breakdown</u> rates, by using stable isotope analysis. With the stable isotope technique, researchers attach a <u>harmless</u> **tracer** to specific amino acids that travel through the bloodstream. Then, they take blood <u>samples</u> to <u>determine</u> the amount of amino acids that enter and exit the leg.

tracer: a material, such as a dye or a radioactive isotope, that can be identified and followed through the course of a process, like a label.

"If 80 amino acids are coming into the artery and 60 are going out of the vein, we know that 20 were probably made into proteins in the muscle," said Dr. Douglas Paddon-Jones, also of UTMB and a co-investigator performing these studies. "We complete the muscle analysis by removing a small piece of muscle and determining how many amino acids have been incorporated into proteins. Over time, we can calculate the rate at which the synthesis and breakdown occurs."

Space conditions also elevate the body's level of the stress hormone cortisol, which increases the breakdown rate of proteins. "Under stress, the body breaks down proteins to make energy for survival," said Ferrando, a member of NSBRI's nutrition and fitness research team. "However, this process also causes muscle atrophy."

To study the supplement's effects on muscle loss due to elevated levels of cortisol, researchers infused the stress hormone into the participants' blood during the stable isotope tests. The researchers mimic the cortisol concentrations found during space flight, then determine protein synthesis and breakdown rates of the subjects taking the supplement and compare this to the rates of the control group.

Ferrando and Wolfe are also collaborating with other NSBRI researchers who use the subjects' body fluids to study changes in bone, immune function and cell damage induced by bed rest.

Findings from this research on nutritional supplements could benefit patients on Earth.

"Muscle atrophy is common in many populations: the elderly, kids with burns, patients in intensive care or people who have had major operations. We're looking at this phenomenon in terms of space flight, but the study has many other implications," Ferrando said.

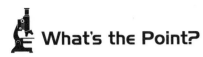

What's the Point?

I. Show your understanding of the reading. Based on the reading, choose the best answer to complete the statements.

1. The subjects in this research experiment on muscle loss are _____.

    a. astronauts

    b. sick patients

    c. medical doctors

    d. people on earth

2. Astronauts in space generally lose muscle mass in _____.

    a. the arms

    b. the back

    c. the heart

    d. the legs

3. People lose muscle when they are not physically active because _____.

    a. only inactive muscles break down

    b. inactive muscles turn into fat

    c. inactive muscles do not produce protein

    d. only active muscles resist disease

4. So far, the study results show that nutritional supplements of amino acids _____.

    a. help increase muscle mass

    b. do not improve muscle health

    c. may cause minor illnesses

    d. help decrease heart rate

5. Researchers determine how much new protein is being produced in the body by _____.

    a. weighing the subjects before and after the experiment

    b. weighing the blood samples before and after the study

    c. counting all amino acids in the body

    d. counting the number of amino acids that enter and exit the leg

6. Muscle atrophy in space is caused not only by inactivity, but also by _____.

    a. poor nutrition

    b. stress

    c. illness

    d. radiation

7. Results of this study _____.

    a. benefit only astronauts in space

    b. are not public information

    c. can apply to patients on earth

    d. are not useful for children

II. Demonstrate your understanding of some of the details in the reading. Based on the text you just read, give short answers to the questions.

1. What are *amino acids?*

2. What is a *placebo?* Based on its use in the reading, can you guess what it is?

3. What is *cortisol?* Where does it come from?

4. Why do the research scientists in this study use a *tracer?*

5. Why do the researchers remove a small piece of muscle?

6. What organization could you contact for more information about medical research in space?

💡 Understanding Words and Phrases

I. Some incomplete sentences follow. Use your own ideas to finish these statements. There is more than one way to complete them. The underlined words can be found in the text you just read.

1. Some ways in which robots can <u>mimic</u> human behavior are

   _____.

2. The <u>confined</u> space of an apartment doesn't allow me _____

   _____.

3. He <u>attempted</u> to remove the red wine stain from his shirt by

   _____.

4. When she was redecorating her house, she compared several <u>samples</u>
   of _____ in order to _____

   _____.

5. In your next essay, please <u>incorporate</u>   _____

   _____.

6. It is often helpful to <u>break down</u> a large problem into smaller parts
   because   _____

   _____.

II. Now, work with a partner to answer the following questions based on your own knowledge. The underlined words appear in the text you just read.

1. How do you <u>determine</u> the area of a rectangle with length L and
   width W?

2. Which <u>harmless</u> animals make good pets?

3. What things in our environment are <u>harmful</u> to your health?

4. What <u>raw</u> materials are needed to build a house?

5. What countries does North America <u>encompass</u>?

6. What are some <u>indications</u> of the flu (influenza)?

7. Why do researchers from different organizations frequently <u>collaborate</u> on a project?

8. What can cause a <u>breakdown</u> in communication between two people?

III. Choose one of the words to fill in the blank in each sentence. You may have to change the form of the word to fit the sentence. Remember, sometimes an English word is both a noun and a verb.

attempt	mimic	sample
breakdown/break down	confinement	synthesis
incorporate	synthesize	

1. Green plants produce sugar through the _____ of carbon dioxide and water.

2. To make more delicious pancakes, add a teaspoon of vanilla to the pancake batter, then stir well to _____ it into the batter.

3. Every year, city officials collect and test _____ of drinking water to make sure that the water is safe to drink.

4. Before 1903, people _____ but failed to achieve heavier-than-air flight. However, on December 17, 1903, the Wright brothers finally succeeded in their _____ to fly a powered airplane.

5. In a process called electrolysis, electric current causes the _____ of water into its elementary components, oxygen and hydrogen.

6. She became seriously ill with pneumonia, which resulted in her _____ to a hospital for a month.

7. Parrots are colorful tropical birds that can _____ human speech.

8. The architect produced the final design for the office building by _____ the ideas of many people.

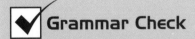 **Grammar Check**

Agent Suffix –er

The article you just read contains the words *researcher* and *tracer*. What do these words mean? How are they related to the words *to research, to trace?*

Grammatical forms are used to convey meanings. In English, the suffix *-er* has several uses and meanings. One common use of *-er* is as an attachment to an adjective to make the comparative form of the adjective: *quick* → *quicker, small* → *smaller, clean* → *cleaner*. Another common use of *-er* is as an ending of a noun. In this case, the suffix has two slightly different meanings.

One meaning of *-er* in a noun is an **agent** or **doer**. Here, the suffix attaches to a verb and changes it to a noun, meaning someone or something that does the action of the verb. So, a *research<u>er</u>* is someone who does research; a *trac<u>er</u>* is something that can trace (or allows others to trace). Also notice the word *do<u>er</u>* in the first sentence of this paragraph!

I. Provide the correct verb for the given agent (in italics).

1. An *interpreter* is someone who _____.

2. A *singer* is a person who _____.

3. A *baker* is a person who _____.

4. An *organizer* is something (or someone) that helps to _____.

5. A *manager* is someone who _____.

6. A *driver* is a person who _____.

7. A *slow cooker* is a special pot that _____.

II. Now, do the reverse: Provide the correct form of the agent noun.

1. Someone who speaks is a _____.

2. A person who plays is a _____.

3. The object used for erasing (chalk or pencil marks) is an _____.

4. Someone who trades is a _____.

5. A machine that computes is a _____.

6. A person who wins is a _____.

7. A pot that cooks under pressure is a *pressure* _____.

III. The other meaning of a noun with the suffix -er is someone who belongs to a group or a place. Here, -er is attached to a noun (a group or place) or to a modifier (which describes the group or place). So, a common*er* *is a person who belongs to the* common *(not noble) class. Some other examples follow. Provide the correct form of the noun and the correct indefinite article (a or an), if needed. In each example, is the suffix attached to a noun or to a modifier?*

1. A person who lives in *New England* is _____.

2. Residents of *New York* are _____.

3. Someone who lives on an *island* is _____.

4. A member of a *troupe* (a group of actors or singers) is

   _____.

5. A soldier who belongs to a *troop* (a military group that uses tanks or horses) is _____.

6. Children who attend summer *camp* are _____.

7. A person who lives in or comes from a *foreign* place is

   _____.

8. Someone who is *inside* a group or organization is _____.

9. People who are *outside* a group or organization are

   _____

10. A person in the *teenage* group (13-19 years old) is _____.

Let's Talk about It

1. Researchers are working on novel techniques for providing health care in space. Such techniques must work in the confined space of a spacecraft. Make some predictions about new medical methods and tools that will be used in space travel. Based on your own logical reasoning, describe some characteristics that special techniques and equipment should have in order to be used successfully in space. If you wish, you may also do some research to answer this question.

2. Can you think of special situations on earth where people might need medical care but don't have easy access to a doctor or to medical equipment?

3. What do people do in these situations?

What Do You Think?

Imagine that you are going away on a year-long pleasure trip on a small ship. You are going with a few friends. How would you prepare for this trip? What things are important to consider? Write a brief plan for your trip. What would you bring with you to take care of your medical needs? Why? What would you like to bring along but can't because there is not enough space? How would you deal with some potential problems? Write a position paper convincing your family, friends, and yourself that this is a good, safe plan.

Expansion Activities

The NSBRI has an Internet website: *www.nsbri.org*. On your own time, visit the website and see what interesting information you can find. You can do this alone or with a classmate. Then, tell your class what you found and what was especially interesting to you.

Note: Most public and university libraries offer free Internet access.

Green Mountain Power's Searsburg Wind Facility.

Wind Power

Search Your Knowledge

1. What's a wind turbine? How does it work? How is it similar to a windmill?

2. How is a wind turbine similar to an electric fan? How are they different?

3. Is wind power used widely around the world? Why or why not?

4. Can you name other sources of energy that are used by power plants to produce electricity?

5. What are the advantages and disadvantages of wind power over other sources of energy?

6. How can the disadvantages be overcome?

The photograph on page 47 shows a currently operating wind-turbine plant. The Searsburg Wind Facility belongs to the Green Mountain Power Corporation, an electric company in Vermont.

You are about to read a text about wind power. The article appeared in the July/August 2002 issue of *Technology Review*, a magazine published by the Massachusetts Institute of Technology (MIT). The reading contains some words and concepts related to wind turbines and power production. Some of these words and ideas are new to many native speakers of English as well. You don't need to understand all the technical details in order to understand the main story. You may wish to skim this article before reading it more thoroughly. (For tips about skimming, see the section called How to Read on page xiii.)

The text contains a number of underlined words. These words may be new to you; therefore, they are underlined so that you can find them easily later on, if you wish to refer to them again. See if you can figure out what they mean from context or from the other words and meanings around the underlined word. The words are also included in the vocabulary exercises under Key Words and Understanding Words and Phrases.

The boldfaced words in the text are glossed in the margin. These non–high frequency vocabulary words or phrases are helpful to understanding the reading.

KEY WORDS

The text in this unit contains the words *blade, prototype, flexible, hinge, niche,* and *capacity.* Do you know what these words mean? If you don't, see if you can figure them out from the sentences that follow. Then, see how the words are used in the article.

1. The <u>blades</u> of a fan, like the blades of a knife, cut through air and make the air move.

2. After the engineers successfully tested the <u>prototype</u> of a newly designed electric car, the managers of the company decided to go ahead with mass production of this design.

3. The Olympic gymnast was very <u>flexible</u>. She could bend backward so that her head touched her feet.

4. When a door squeaks, it's time to oil its <u>hinges</u> to make it move more smoothly.

5. a. A mouse found a <u>niche</u> in the side of a hill and made a nest there.
   b. The little diner that serves only soup found a small but successful <u>niche</u> in the large restaurant market.

6. a. The gas tank in my car has a <u>capacity</u> of 16 gallons.
   b. When the candy factory operates at full <u>capacity</u>, it produces 100 chocolate bars an hour.

Wind Power for Pennies

A Lightweight Wind Turbine Is Finally on the Horizon—and It Might Just Be the <u>Breakthrough</u> Needed to Give Fuels a Run for Their Money

Peter Fairley
from *Technology Review*

on the horizon (idiom): going to happen soon

give fuels a run for their money (idiom): give them strong competition (*a run for one's money*)

proving ground (idiom): a place for testing new devices or theories

The newest wind turbine standing at Rocky Flats in Colorado, the U.S. Department of Energy's **proving ground** for wind power technologies, looks much like any other apparatus for capturing energy from wind: a boxy turbine sits atop a steel tower that sprouts two propeller <u>blades</u> stretching a combined 40 meters—almost half the length of a football field. Wind rushes by, <u>blades</u> rotate, and electricity flows. But there's a key difference. This <u>prototype</u> has <u>flexible</u>, <u>hinged</u> <u>blades</u>; in strong winds, they bend back slightly while spinning. The bending is barely perceptible to a casual observer, but it's a radical <u>departure</u> from how existing wind turbines work—and it just may change the fate of wind power.

Indeed, the success of the <u>prototype</u> at Rocky Flats comes at a <u>crucial</u> moment in the evolution of wind power. Wind-driven generators are still a <u>niche</u> technology—producing less than one percent of U.S. electricity. But last year, 1,700 megawatts' worth of new wind <u>capacity</u> was installed in the United States—enough to power 500,000 houses—nearly doubling the nation's wind power <u>capacity</u>. And more is on the way. Manufacturers have reduced the cost of heavy-duty wind turbines fourfold since 1980, and these gargantuan machines are now reliable and efficient enough to be built offshore. An 80-turbine, $240 million wind farm under construction off the Danish coast will be the world's largest, and developers are beginning to colonize German, Dutch and British waters, too. In North America, speculators <u>envision</u> massive offshore wind farms near British Columbia and Nantucket, MA.

But there is still **a black cloud** hovering over this seemingly sunny scenario. Wind turbines remain expensive to build—often prohibitively so. On average, it costs about $1 million per megawatt to construct a wind turbine farm, compared to about $600,000 per megawatt for a <u>conventional</u> gas-fired power plant; in the economic calculations of power companies, the fact that wind is free doesn't close this gap. In short, the price of building wind power must come down if it's ever to be more than a <u>niche</u> technology.

a black cloud (idiom): a sign of misfortune; bad luck

And that's where the <u>prototype</u> at Rocky Flats comes in. The flexibility in its blades will enable the turbine to be 40 percent lighter than today's industry standard but just as capable of surviving destructive storms. And that lighter weight could mean machines that are 20 to 25 percent cheaper than today's large turbines.

Earlier efforts at lighter designs were <u>universal</u> failures—disabled or destroyed, some within weeks, by the wind itself. Given these failures, wind experts are understandably cautious about the latest shot at a lightweight design. But most agree that lightweight wind turbines, if they work, will change the economic equation. "The question would become, 'How do you get the transmission <u>capacity</u> built fast enough to keep up with growth,'" says Ward Marshall, a wind power developer at Columbus, OH-based American Electric Power who is on the board of directors of the American Wind Energy Association, a trade group. "You'd have plenty of folks willing to sign up."

And, say experts, the Rocky Flats <u>prototype</u>—designed by Wind Turbine of Bellevue, WA—is the best hope in years for a lightweight design that will finally succeed. "I can say pretty <u>unequivocally</u> that this is a dramatic step in lightweight [wind turbine] technology," says Bob Thresher, director of the National Wind Technology Center at Rocky Flats. "Nobody else has built a machine that <u>flexible</u> and made it work."

Steady as She Blows

Wind turbines are like giant fans run in reverse. Instead of motor-driven blades that push the air, they use airfoils that catch the wind and crank a generator that pumps out electricity. Many of today's turbines are mammoth machines with three-bladed rotors that span 80 meters—20 meters longer than the wingspan of a Boeing 747. And therein lies the technology challenge. The enormous size is needed if commercial wind turbines are to compete economically because power production rises exponentially with blade length. But these vast structures must be **rugged** enough to endure gales and extreme turbulence.

rugged: strong

In the 1970s and '80s, U.S. wind energy pioneers made the first serious efforts at fighting these forces with lightweight, <u>flexible</u> machines. Several startups installed thousands of such wind turbines; most were literally torn apart or disabled by **gusts.** Taking lightweight experimentation to the extreme, General Electric and Boeing built much larger <u>prototypes</u>—behemoths with 80-, 90-, and even 100-meter-long blades. These also proved <u>prone</u> to breakdown; in some cases their blades bent back and actually struck the towers.

gusts: sudden and violent rushes of wind

All told, U.S. companies and the Department of Energy spent hundreds of millions of dollars on these failed experiments in the 1980s and early 1990s. "The American model has always gravitated toward the light and sophisticated and things that didn't work," says James Manwell, a mechanical engineer who leads the University of Massachusetts's renewable-energy research laboratory in Amherst, MA.

Into these technology doldrums sailed researchers from Denmark's Risø National Laboratory and Danish companies like Vestas Wind Systems. During the past two decades they perfected a heavy-duty version of the wind turbine—and it has become the Microsoft Windows of the wind power industry. Today, this Danish design accounts for virtually all of the electricity generated by the wind worldwide. Perhaps reflecting national inclinations, these sturdy Danish designs had little of the aerodynamic flash of the earlier U.S. wind turbines; they were simply braced against the wind with heavier, thicker steel and composite materials. They were tough, rugged—and they worked.

What's more, in recent years, power electronics—digital silicon switches that massage the flow of electricity from the machine—further improved the basic design. Previously, the turbine's rotor was held to a constant rate of rotation so its alternating-current output would be <u>in sync</u> with the power grid; the new devices maintain the <u>synchronization</u> while allowing the rotor to freely speed up and slow down with the wind. "If you get a gust, the rotor can accelerate instead of just sitting there and receiving the brute force of the wind," says Manwell.

Mastering such strains enabled the Danish design to grow larger and larger. Whereas in the early 1980s a typical commercial machine had a blade span of 12.5 meters and could produce 50 kilowatts—enough for about a dozen homes—today's biggest blades stretch 80 meters and **crank out** two megawatts; a single machine can power more than 500 homes.

crank out: produce, especially mechanically

The newest challenge facing the Danish design is finding ways for it <u>to weather</u> the corrosive and punishing offshore environment, where months can pass before a mechanic can safely board and fix a turbine. Vestas, for one, is equipping its turbines with sensors on each of their components to detect **wear and tear**, and backup systems to take over in the case of, say, a failure in the power electronics.

wear and tear (idiom): damage resulting from normal use or exposure

Vestas's approach goes to the test this summer, as Denmark's power supplier begins installing 80 Vestas machines in shallow water 14 to 20 kilometers off the Danish coastline. It will be the world's biggest offshore wind farm, powering as many as 150,000 Danish households.

Wind Shadows

These upgrades will make big, heavy turbines more reliable, but they don't add up to a fundamental <u>shift</u> in the economics of wind power. Nations like Denmark and Germany are prepared to pay for wind power partly because fossil fuels are so much more costly in Europe, where higher taxes cover environmental and health costs associated with burning them. (About 20 percent of Denmark's power comes from wind.) But for wind power to be truly cost competitive with fossil fuels in the United States, the technology must change.

What makes Wind Turbine's Rocky Flats design such a <u>departure</u> is not only its <u>hinged</u> blades, but also their downwind orientation. The Danish design faces the blades into the wind and makes the blades heavy so they won't bend back and slam into the tower. The Wind Turbine design can't face the wind—the <u>hinged</u> blades would hit the tower—so the rotor is positioned downwind. Finally, it uses two blades, rather than the three in the traditional design, to further reduce weight.

A LIGHTER, CHEAPER TURBINE

Hinged blades and sophisticated control systems allow the lightweight turbine designed by Wind Turbine of Bellevue, WA, to survive storms and gusts.

In Normal Conditions: Blades spin freely, the entire turbine swivels according to wind direction, and a gearbox amplifies blade rotation speed so a generator can produce power.

In High Wind or Erratic Conditions: Hydraulic dampers allow blades to flex up to 15 degrees downwind and five degrees upwind to shed excess wind force. Control systems include a brake to slow blade spin and a yaw drive to counteract swiveling.

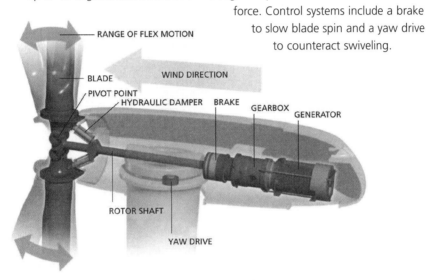

RANGE OF FLEX MOTION

BLADE

WIND DIRECTION

PIVOT POINT
HYDRAULIC DAMPER   BRAKE   GEARBOX   GENERATOR

ROTOR SHAFT

YAW DRIVE

Advances in the computer modeling of such dangerous forces as vibration helped the design's development. <u>Flexible</u> blades add an extra dimension to the machine's motion; so does the fact that the whole machine can freely swivel with the wind. (Traditional designs are driven to face the wind, then locked in place.) Predicting, detecting and preventing disasters—like rapidly <u>shifting</u> winds that swing a rotor upwind and send its <u>flexible</u> blades into the tower—are control challenges even with the best design. "If you don't get that right, the machine can literally beat itself to death," says Ken Deering, Wind Turbine's vice president of engineering.

Two years ago, when Wind Turbine's <u>prototype</u> was erected at Rocky Flats, there were worries that this machine, too, would beat itself to death. Thresher says some of his staff feared that the machine, like its 1980s predecessors, would not long escape **the scrap heap.** Today, despite some minor setbacks, those doubts are fading.

the scrap heap (idiom): a place for discarding useless material

Emboldened by its early success, Wind Turbine has installed, near Lancaster, CA, a second prototype, with a larger, 48-meter blade span. By the end of this year, the company expects to boost blade length on this machine to 60 meters—full commercial size. What's more, this new prototype has a thinner tower, aimed at reducing the noisy thump—known as a "wind shadow"—that can occur each time a blade whips through the area of turbulent air behind the tower. And with its lighter weight, the turbine can be mounted atop higher towers, reaching up to faster winds.

Becalmed

Whatever the advances in technology, however, the wind power industry still faces significant hurdles, starting with uncertain political support in the United States. In Europe, wind power is already a relatively easy sell. But in the United States, wind developers rely on federal tax credits to make a profit. These vital credits face chronic opposition from powerful oil and coal lobbies and often lapse. The wind power industry raced to plug in its turbines before these credits expired at the end of last year, then went dormant for the three months it took the U.S. Congress to renew them. Congress extended the credits through the end of next year, initiating what is likely to be yet another start-and-stop development cycle.

A second obstacle to broad adoption is the wind itself. It may be free and widely accessible, but it is also frustratingly inconsistent. Just ask any sailor. And this **fickleness** translates into intermittent power production. The more turbines get built, the more their intermittency will complicate the planning and management of large flows of power across regional and national power grids. Indeed, in west Texas, a recent boom in wind turbine construction is straining the region's transmission lines—and also producing power out of sync with local needs: wind blows during cool nights and stalls on hot days when people most need electricity.

fickleness: unpredictability or instability, especially of feelings; unreliability

Texas utilities are patching the problems by expanding transmission lines. But to really capture the value of wind power on a large scale, new approaches are needed to storing wind power when it's produced and releasing it when needed. The Electric Power Research Institute, a utility-funded R&D consortium in Palo Alto, CA, is conducting research on how to make better one-day-ahead wind predictions. More important, it is exploring ways to store energy when the wind is blowing. "We need to think about operating an electrical system rather than just focusing on the wind turbines," says Chuck McGowin, manager for wind power technology at the institute. Storage facilities "would allow us to use what we have more efficiently, improve the value of it."

In the northwest United States, one storage option being developed by the Portland, OR-based Bonneville Power Administration balances wind power with hydroelectric power. The idea is simple: when the wind is blowing, don't let the water pass through the hydroelectric turbines; on calm days, open up the gates. And the Tennessee Valley Authority is even experimenting with storing energy in giant fuel cells; a pilot plant is under construction in Mississippi.

Wind power faces plenty of obstacles, but there's more reason than ever to believe these obstacles will be overcome. Worries over the environmental effects of burning fossil fuels and political concerns about an overdependence on petroleum are spurring a boom in wind turbine construction. But it is advances in technology itself, created by continued strong research efforts, that could provide the most critical impetus for increased use of wind power.

At Rocky Flats, four rows of research turbines—a total of a dozen machines ranging from 400-watt battery chargers to grid-ready 600-kilowatt machines—share a boulder-strewn 115-hectare plain. With the Rocky Mountains as a backdrop, their blades whip against the breezes blowing in from El Dorado Canyon to the west. At least, they do much of the time. "We have a lot of calm days, in the summer in particular, and for a testing site it's good to have a mix," Thresher says.

Calm days may be good for wind turbine research, but they're still among the biggest concerns haunting wind turbine commercialization. While no technology can make the wind blow, lower-cost, reliable technologies appear ready to take on its fickleness. And that could mean a wind turbine will soon sprout atop a breezy hill near you.

Major Players in Wind Power R&D	
Organization	**R&D Focus**
Bergey WindPower (Norman, OK)	Small (one-kilowatt to 50-kilowatt) turbines for distributed-power applications
General Electric Wind Energy (Atlanta, GA)	Improving existing large, heavy design
Mitsubishi Heavy Industries (Tokyo, Japan)	Large turbines for offshore use
National Wind Technology Center (Rocky Flats, CO)	Cost-effective turbines for moderate-wind sites
Pfleiderer Wind Energy (Neumarkt, Germany)	Simplifying gearbox with permanent-magnet generators
Risø National Laboratory (Ringkobing, Denmark)	Improving existing large, heavy design; testing of small-scale, lightweight designs
Vestas Wind Systems (Roskilde, Denmark)	Improving existing large, heavy design
Wind Turbine (Bellevue, WA)	Large, lightweight flexible turbines

Source: Peter Fairley, "Wind Power for Pennies," *Technology Review,* July/August 2000, copyright © 2002 Technology Review.

What's the Point?

I. Show your understanding of the reading. Based on the reading, choose the best answer to complete these sentences.

1. The new wind turbine being tested at Rocky Flats looks like
   _____.
   a. a normal wind turbine
   b. a football field
   c. a jet airplane turbine

2. Since 1980, wind turbines have become four times _____.
   a. more efficient
   b. less expensive
   c. smaller

3. The most important innovation of the new wind turbine is
   _____.
   a. flexible, hinged blades that face away from the wind
   b. 2 blades instead of the traditional 3
   c. greater height than that of exisitng turbines

4. According to the article, the new turbines would cost less because
   of _____.
   a. cheaper materials
   b. lower weight
   c. lower manufacturing cost

5. Early light-weight wind turbines _____.
   a. were successful in Denmark
   b. were used offshore
   c. all failed

6. The new wind turbine prototype was designed by _____.
   a. the Unites States Department of Energy
   b. the National Wind Technology Center, a government
       agency
   c. Wind Turbine, a company in Bellevue, Washington

7. Early American designs failed because they were _____.

    a. too simple

    b. too high-tech

    c. too expensive

8. Danish designs were successful because the wind turbines were _____.

    a. supported by strong, heavy structures

    b. designed using advanced aerodynamics

    c. constructed from new materials

9. Two important technological advances that made modern turbine designs possible are _____.

    a. power electronics devices and computer modeling

    b. computer modeling and weather forecasting

    c. weather forecasting and manufacturing

10. Since the early 1980s, the power production of a single wind turbine has _____.

    a. not increased significantly

    b. increased four times

    c. increased 40 times

11. The new wind turbine could be installed on a higher tower than before, and could thus reach faster winds, because the tower would _____.

    a. be thicker and, therefore, stronger

    b. be made of new stronger materials

    c. support a lighter turbine

12. For wind power to become more widely accepted and used, it will be necessary _____.

    a. to store wind energy

    b. to replace hydroelectric power plants

    c. to build all wind turbines off shore

II. Complete the sentences below based on the article you just read. To complete a sentence, sometimes you may need a single word and sometimes a phrase. You may use your own words or words from the text.

1. According to the chart in the article, the number of countries that are "major players" in wind power is _____.

2. An important drawback (negative side) of wind power is

   _____.

3. When this article was written, the largest wind power plant was in

   _____.

4. Wind turbines must be very large in order to _____

   _____.

5. In the United States, the most abundant and least expensive form of energy for producing electricity is _____.

6. In Europe, fossil fuels are very costly because of _____

   _____.

7. A commercial wind-turbine machine built by Wind Turbine Company will have a combined blade length of _____.

8. A power-plant wind turbine can produce as much power as _____ kilowatts.

💡 Understanding Words and Phrases

I. Choose one of the words to fill in the blank in each sentence. You may have to change the form of the word to fit the sentence.

breakthrough	intermittent	synchronization
conventional	in sync; out of sync	unequivocally
crucial	prone	universal
departure	to shift	to weather
to envision		

1. People's tastes for different kinds of food vary from culture to culture, but their pleasure in eating is _____.

2. Tom woke up many times during the night and felt tired the next day because of his _____ sleep.

3. Ten years ago, a great fire destroyed most of the family's farm, but they _____ this catastrophe and are still running the farm today.

4. When Diane heard her name mentioned by her colleagues, she _____ her attention from her work to their conversation.

5. Gandhi's peaceful protest against the government of Great Britain was a _____ from the more _____ violent rebellions against ruling regimes.

6. Katie was poor and had a difficult life, but she worked very hard and _____ a brighter future for her two children.

7. After the dog lost one of its legs in an accident, it was _____ to falling down if it ran too fast.

8. Helen's office clock, wristwatch, and computer clock are all _____, so she is never sure of the correct time.

9. For a while, the factory workers weren't sure whether or not they would lose their jobs. But, in a recent announcement, the company president stated _____ that the factory would shut down in six months and all employees would be laid off.

10. For a psychologist, the ability to listen is _____.

11. The discovery of antibiotics was a major _____ in treating infections.

12. For our home movie, we recorded the action on a videotape and the sound on an audiotape. When we showed the movie, we had to _____ the two tapes to match the action with the sound.

II. The article uses several idiomatic expressions. They are defined in the text margins. Each of these phrases is used as a single unit that has a particular meaning. Review the use of these expressions in the article. Then, select the phrase that fits best in each of the following sentences.

on the horizon	a proving ground	a black cloud
wear and tear	the scrap heap	give (someone) a run for (one's) money

1. My little 40-year-old refrigerator is not ready for
   _____ yet; it still runs very well and
   keeps the food cold.

2. Many older people suffer from arthritis due to normal
   _____ on their joints.

3. The Olympic training camp is _____ for
   the country's best athletes. The few who succeed there will go on to
   compete in the Olympic Games.

4. The basketball coach said, "The other team is very strong, but we
   can beat them. At least, we'll _____!"

5. When Laura finally began to recover from her long and dangerous
   illness, _____ was lifted from her family,
   and they celebrated the good news.

6. Yes, their wedding is _____; they'll
   probably be married before the end of the year.

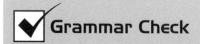
Grammar Check

Adverb Phrases

The article you just read contains these sentences:

1. ***Taking*** *lightweight experimentation to the extreme,* **General Electric and Boeing** built much larger prototypes.

2. *Perhaps **reflecting** national inclinations,* these sturdy Danish **designs** had little of the aerodynamic flash of the earlier U.S. wind turbines.

The part of the sentence that is in *italics* is called an **adverbl phrase**, or **participial phrase.** An adverb phrase modifies the subject of the sentence. Notice that the verb in the adverb phrase is in the form *-ing*, which is also called the *present participle*; this gives the name *participial* phrase. The adverb phrase comes from reducing, or changing, an *adverb clause* (see Grammar Check: Adverb Clauses in Unit 2). To illustrate this, let's analyze the first example above.

Sentence with adverb clause:

> ***Because they took*** *lightweight experimentation to the extreme,* **General Electric and Boeing** built much larger prototypes.

Sentence with modifying adverb phrase:

> ***Taking*** *lightweight experimentation to the extreme,* **General Electric and Boeing** built much larger prototypes.

Notice that the subject of the adverb clause (***they***) and the subject of the main clause (**General Electric and Boeing**) refer to the same thing. The modifying adverb phrase modifies the subject **General Electric and Boeing**. In the adverb phrase, the subject is omitted, and the verb takes the *-ing* form; the conjunction *because* is not used.

Can you rewrite the second example from the article so that it contains an adverb clause?

Rules of reduction:

1. Omit the subject of the adverb clause.

2. Change the verb in the adverb clause to the *-ing* form. If the verb is in the progressive tense, *be + -ing*, then omit the *be* form and just use the *-ing* verb. For example:

   > While **she was listening** to music, **she** cleaned the apartment.
   >
   > → While **listening** to music, **she** cleaned the apartment.

3. Time phrases: Modifying adverb phrases can be formed by reducing adverb clauses that begin with the conjunctions *after, before, since, when,* and *while*. The adverb phrase may come before or after the main clause. When the adverb phrase is at the end of the sentence, there is no comma. For example:

   > **Since I met** you, I have learned a lot about Canada.
   >
   > → **Since meeting** you, I have learned a lot about Canada.
   >
   > → I have learned a lot about Canada **since meeting** you.

Sometimes, the conjunction **while** can be omitted, but then the adverb *-ing* phrase must come at the beginning of the sentence. For example:

> **While he waited** for the bus, **Tom** watched the old man.
>
> → **While waiting** for the bus, **Tom** watched the old man.
>
> → **Waiting** for the bus, **Tom** watched the old man.
>
> Incorrect (different meaning!): Tom watched the old man **waiting** for the bus.
>
> Tom watched the old man **while** waiting for the bus.

4. Cause-and-effect phrases: An adverb phrase at the beginning of a sentence often has the meaning "because," even though the conjunction *because* is always omitted. For example:

    a. **Because she was sick** *with the flu*, Renee couldn't go to work.

      → **Being sick** *with the flu*, Renee couldn't go to work.

      → **Sick** *with the flu*, Renee couldn't go to work.

    b. **Because she had eaten** *a big lunch*, Kathy wasn't hungry for dinner.

      → **Having eaten** *a big lunch*, Kathy wasn't hungry for dinner.

In cause-and-effect sentences, the adverb phrase must come before the main clause. Notice in example *a* that a form of *be* may be changed to **being** or it may be omitted. In example *b*, the use of the perfect tense, **having** + **PAST PARTICIPLE,** means "because + before (an earlier action)."

5. Important: An adverb clause can be reduced to an adverb phrase only when the subject of an adverb clause is the same as the subject of the main clause. If the subjects are different, a reduction is not possible. For example:

    a. *Because* **Jerry** *wanted to treat us*, **we** let him buy ice cream.

    Incorrect: *Wanting to treat us*, **we** let him buy ice cream.

    b. *While* **the music** *was playing*, **Anna** prepared lunch.

    Incorrect: *Playing*, **Anna** prepared lunch.

In these examples, the incorrect phrases *Wanting to treat us* and *Playing* are called *dangling participles*—they are "hanging alone" because there is no correct subject to modify.

Exercise 1

In each sentence, change the adverb clause to an adverb phrase. What is the subject of the adverb clause? What is the subject of the main clause? Which sentences must begin with the adverb phrase? Which sentences cannot be changed?

1. While Jeremy was taking a shower, he sang his favorite songs.

2. Because previous methods were unsuccessful, the mathematician tried something new.

3. Always look both ways when you cross the street.

4. The lawyers prepared their case very carefully before they defended their client in court.

5. After he had finished his second piece of apple pie, Clarence took a third piece.

6. Because I've never been to Hawaii before, I really look forward to going there this winter.

7. Peter has been taking care of his mother at home since she became ill.

8. Sally is frequently invited to parties because she can perform magic tricks.

Exercise 2

Think about the meanings of the following sentences. Which ones express the idea of time, or cause and effect, or perhaps both? Some sentences contain dangling participles; circle the dangling participles, and change the incorrect sentences in some possible way to make them correct.

1. When giving advice, Doreen tries to be polite and tactful.

2. Having been married and divorced five times, Mrs. Post is an expert on weddings but perhaps not on marriage!

3. Being able to explain concepts very clearly, all the students want to be in Professor Moore's science class.

4. Walking on an icy sidewalk, Linda fell and broke her wrist.

5. After arriving home with the suitcases, Charlie realized that they weren't his!

6. While raining outside, the birds hid under the bridge.

7. Wanting to stay dry in the rain, the birds hid under the bridge.

8. Chewing a piece of bread, Martha bit on a small stone and broke her tooth.

9. Since visiting Barbara a year ago, she [Barbara] hasn't called or written to Ben.

10. Unaccustomed to taking orders from others, Rick quit his job at the big company as soon as he could start his own business.

11. Working on a long and challenging homework assignment, Norman ate an entire box of cookies.

12. Traveling through rural New England, we saw many small, family-run dairy farms.

Exercise 3

In each item, combine the given sentences into a single sentence by using one of the sentences to form an adverb phrase. What relationship is there between the adverb phrase and the main clause—time, while, because (cause and effect)? Which items cannot be combined in this way without changing the sentences?

1. Jim was taking a stroll in the city. He ran into two old acquaintances.

2. The book should be discarded from the library. It has several torn and missing pages.

3. The man was crossing the street with two small children. He held onto them very tightly.

4. Maggie was talking to herself on the bus. Other passengers were staring at her.

5. Albert Einstein won a Nobel Prize. He had made significant discoveries in physics.

6. Cats make good house pets. They are small, clean, and unaggressive.

7. Jacob (had) retired from his job. Then, he became even busier with his hobbies and volunteer work.

8. He is a highly respected member of the community. Many people seek his advice.

Let's Talk about It

1. Why is wind power a niche technology? What is the niche—who uses wind power?

2. What are the innovations of this design? How are these features an improvement over existing designs?

3. Why is wind power not widespread in the United States? Do you think the new wind turbine will help?

4. What source or sources of energy are the best for a given country? Explain and defend your answer.

5. The article suggests that the increased use of wind power will depend on economic factors, political concerns, environmental concerns, and technological advances. Which of these do you think plays the most important role? Why?

What Do You Think?

What source of power is the best for your region or community? Imagine that you are part of a team that must recommend a particular power plant, or plants, for your local area. What will you recommend? How will you decide what option is best? Write a letter to your local government office describing your proposal and explaining your reasons for it.

A pygmy horned lizard.

High-Tech Lizards

Search Your Knowledge

1. What is a lizard? What is a gecko?

2. Have you ever seen a lizard or gecko walk on walls or ceilings?

3. How do you think they are able to do this without falling?

4. What is a molecule? What is an electron?

5. Are you familiar with any forces between molecules or electrons that allow materials to stick together?

The two magazine articles that follow talk about the science behind geckos' ability to walk vertically and upside down. The first article explains how geckos' feet are able to stick to various surfaces. The second article describes a new technological invention based on this science. These articles appeared in two different issues of the weekly magazine *Science News*. Can you understand the underlined words in the texts? These words may be new to you; therefore, they are underlined so that you can find them easily later on, if you wish to refer to them again. See if you can figure out what they mean from context or from the other words and meanings around the underlined word. The words are also included in the vocabulary exercises under Key Words and Understanding Words and Phrases.

The boldfaced words in the text are glossed in the margin. These non–high frequency vocabulary words or phrases are helpful to understanding the reading.

First, read Part 1 and complete the exercises for that part. Then, read Part 2 and complete the exercises for that part. Finally, complete the discussion and writing exercises that follow; they pertain to Parts 1 and 2 together.

KEY WORDS

This word is important for understanding the text that you are about to read. It is underlined in the text for easy reference. Study its meanings and answer the related questions.

adhesive: 1. (adjective) sticky. 2. (noun) a material that makes things stick.

Can you give some examples of adhesive materials?
What are the meanings of the verb *adhere* and the noun *adhesion*?

PART 1

Getting a Grip

How Gecko Toes Stick

from *Science News*

Geckos are the <u>envy</u> of rock climbers. Without glue, suction, or claws, these lizards <u>scamper</u> up walls and hang from ceilings.

Scientists finally have <u>pinned down</u> the molecular basis of this seeming magic. Gecko feet are covered by billions of tiny hair tips, or spatulae, that hug surfaces. Temporary shifting of the electrons in the molecules of the spatulae and of opposing rocks, walls, or ceilings creates <u>adhesive</u> van der Waals forces, according to a study in the Aug. 27 Proceedings of the National Academy of Sciences. The collective action of these <u>subtle</u> intermolecular interactions <u>contributes</u> to <u>countless</u> <u>properties</u>, including a liquid's boiling point and a **polymer**'s strength.

Previous research had shown that gecko adhesion relies on intermolecular forces, but scientists weren't sure whether van der Waals <u>bonding</u> or water adsorption was at work. In water adsorption, a thin layer of the liquid acts like glue, but only on surfaces that <u>readily</u> <u>bond</u> water. The new study, however, shows that geckos <u>cling</u> equally well to water-attracting and water-<u>repelling</u> surfaces. Using mathematical models, the authors report that the width of each spatula is just what would be expected if van der Waals forces were operating.

The small size and high density of the spatulae, rather than their chemical composition, <u>enable</u> geckos to stick to the world so well, report Kellar Autumn of Lewis and Clark College in Portland, Ore., and his colleagues.

Gecko spatulae are made of keratin, the protein in human hair. However, when the scientists made spatulae **mockups** out of either **silicon rubber** or **polyester,** each material adhered to many surfaces as well as the real spatulae did.

"Just by <u>splitting</u> a surface into multiple small tips, we can get dry adhesion," Autumn says. Such structures might serve as a new type of <u>adhesive</u> that doesn't require messy, smelly liquids.

The work shows that strong adhesion can arise from what are thought to be relatively weak forces, comments Matthew Tirrell of

polymer: a natural or synthetic (man-made) material, having long chains of molecules. Plastics and rubber are polymers.

mockup: a model of a structure, usually full-sized, used for study, demonstration, or testing

silicon rubber: a water-resistant synthetic material that contains the element silicon

polyester: light and strong synthetic polymer. Some clothes are made of polyester.

the University of California, Santa Barbara. The highly divided gecko foot is also minutely <u>adaptable</u> to bumpy surfaces and is easy to reposition, he says.

A good <u>adhesive</u> has to both stick and release easily, adds Anthony Russell of the University of Calgary in Alberta. "Getting something to stick is not that hard," he notes. "Getting it off and being able to use it again, that is one of the **neat** things that geckos have been able to do."

neat (slang): wonderful, terrific

What's the Point?

Check your understanding of the text. Based on the reading, are the following statements true (T) or false (F)?

1. _____ Geckos are able to walk on ceilings due to special intermolecular forces.

2. _____ Van der Waals adhesion requires a thin layer of water.

3. _____ A gecko's feet can stick to any surface because of their special chemical composition.

4. _____ Artificial spatulae adhere to surfaces as well as natural gecko spatulae.

5. _____ The size of the spatulae on a gecko's foot is not important for adhesion.

6. _____ The number of spatulae on a gecko's foot is very important for adhesion.

7. _____ Van der Waals forces are among the strongest forces in nature.

8. _____ The most amazing thing about the adhesive property of gecko feet is that they can stick to bumpy surfaces.

Understanding Words and Phrases

I. Each item in the left column represents a category, and each item in the right column give examples of things that belong in one of the categories. Match each item on the left with the correct examples on the right. Words that are underlined here are also underlined in the text for your reference.

_____ 1. things that are <u>countless</u>	a. glue; cement
_____ 2. things that <u>bond</u>	b. legs; wheelchair; airplane
_____ 3. things that <u>repel</u> other things	c. stars in the sky; grains of sand on the beach
_____ 4. things that <u>enable</u> us to move	d. insect spray; magnets with like poles

Can you add other examples to each category in the left column?

II. Words that have the same or similar meanings are synonyms. For each word or phrase in the left column, pick the correct synonym in the right column. The words on the left are underlined in the text for easy reference. If a word has more than one meaning, use the meaning that is used in the article.

_____ 1. envy	a. easily
_____ 2. scamper	b. establish exactly
_____ 3. pin down	c. jealous desire
_____ 4. subtle	d. divide
_____ 5. contribute	e. fit
_____ 6. readily	f. run lightly
_____ 7. cling	g. share in the cause
_____ 8. split	h. delicate
_____ 9. adapt	i. characteristic or quality
_____ 10. property	j. hold on tightly

III. Answer the following questions or do the requested tasks. The underlined words are also underlined in the article for easy reference. Respond in full sentences.

1. What is something that you <u>envy</u>?

2. Give an example of something or someone that <u>scampers</u>.

3. Name something that is difficult to <u>pin down</u>.

4. What is a <u>subtle</u> way to get someone's attention?

5. What <u>contributes</u> to air pollution?

6. Name something that is <u>readily</u> available to you.

7. Give an example of someone or something that <u>clings</u> to another person or thing.

8. Describe some things that can be <u>split</u>.

9. What circumstances have you found difficult to <u>adapt</u> to? What has been easy to <u>adapt</u> to?

10. What are some <u>properties</u> of gold?

✔ Grammar Check

Present Perfect and Past Perfect Tenses

The second paragraph of the text begins with the sentence "Scientists finally **have pinned down** the molecular basis of this seeming magic." The words in bold type are the ***present perfect*** tense of the verb *pin down*. The form of the present perfect is: **have/has + PAST PARTICIPLE**. What is the difference between this present perfect form and the simple past form *pinned down*?

The simple past tense means that an event happened at a specified time in the past and is finished; it was completed in the past. The perfect tense means that one event happens *prior to* another event or time; the prior event may not be completed and its time is not specified. The perfect tense often implies duration or repetition of an event prior to a subsequent event.

The present perfect tense means that an event happened at some unspecified time prior to the present. It also gives the idea that this event is still relevant now. So, as stated in the article, scientists *have pinned down* the explanation for geckos' magical ability—at some unspecified time before now—and this explanation is still true today.

The third paragraph of the article begins with the sentence, "Previous research **had shown** that gecko adhesion relies on intermolecular forces . . ." The words in bold type are the ***past perfect*** tense of the verb *pin down*. Notice the form of the past perfect: **had + PAST PARTICIPLE**. The past perfect tense means that an event happened in the past prior to another event or time in the past. So, the article says that previous research *had shown* something at some unspecified time prior to the new research and was still relevant when the new research was done; both research events (previous and new) took place in the past.

Compare the following examples:

Simple past
Last year, the chef **created** a new cake recipe, which she included in her latest cookbook.

Present perfect
She **has created** a new soup recipe, which is very popular at her restaurant today.

Past perfect
She **had created** many other recipes before she created the new cake recipe.

I. Complete the following sentences by choosing the correct tense of the verb in parentheses: **simple past, present perfect,** *or* **past perfect.**

1. Last fall, the children (*go*) _____ to an art museum. They (*be*) _____ to a science museum before, but (*be*, never) _____ to an art museum.

2. My parents (*visit*) _____ art and history museums many times.

3. We finally (*go*) _____ hiking yesterday. We (*hike*, not) _____ in several months.

4. My teacher (*move*) _____ to Vermont in 1970 and (*live*) _____ there ever since.

5. By the time Mark and Sofia got married, they (*know*) _____ each other six years.

6. No, thank you, I'm not thirsty. I (*have*, already) _____ two glasses of juice.

7. So far, he (*complete*) _____ three research projects but (*write*, not) _____ the reports yet.

8. When we entered the laboratory, the experiment (*start*, just) _____.

9. Last night, my brother (*eat*) _____
   an entire banana split. He (*be able*) _____
   to finish it because he (*have*) _____ nothing
   else to eat that day.

10. Europeans (*taste*, not) _____ chocolate
    until they were introduced to it by Native Americans.

11. Scientists still (*figure out*, not) _____ all
    the secrets of our solar system.

12. She (*feel*) _____ embarrassed when
    she arrived late for the conference. She (*realize*, not)
    _____ that people were waiting for her.

*II. According to the answers in Part I, what tenses are usually used with
the following expressions: ever, never, already, just, still, yet, many times,
so far, by the time? For each of these expressions, write your own sentence
using that expression.*

1. ever: _____

2. never: _____

3. already: _____

4. just: _____

5. still: _____

6. yet: _____

7. many times: _____

8. so far: _____

9. by the time: _____

PART 2

Caught on Tape

Gecko-inspired adhesive is superstrong

from *Science News*

As it <u>scurries</u> along the ceiling, a gecko has the sticking power to support not just its own body weight, but about 400 times as much. Besides that sticking power, the natural adhesive on this animal's feet is clean and reusable, and it works on all surfaces, wet or dry.

Scientists at the University of Manchester in England and the Institute for Microelectronics Technology in Russia have <u>emulated</u> the animal's adhesive mechanism by creating "gecko tape." It comes closer to the lizard's sticking power than any other gecko-styled adhesive so far.

The 1-square-centimeter prototype patch can <u>bear</u> about 3 kilograms, almost one-third the weight that the same area of gecko sole can support.

In the July *Nature Materials*, Andre Geim of the University of Manchester and his <u>colleagues</u> <u>claim</u> that the tape is scalable to human dimensions: Wearing a "gecko glove," a person could <u>dangle</u> from the ceiling. In theory, the tape could hold **tissues** together after surgery or support stunt doubles climbing around movie sets.

tissues (biology): collections of cells

The gecko tape is modeled on the gecko sole, an <u>intricate</u> fingernail-size surface covered with a half-million microscopic, hair-like structures known as setae. Each seta's tip <u>branches</u> into even **finer** hairs that <u>nestle</u> so closely with every surface the gecko touches that intermolecular attractions called van der Waals bonds and capillary forces **kick in.** These bond the gecko's foot to the surface.

finer: thinner (adjective: *fine*)

kick in (informal): begin to take effect

tidy: orderly and neat

Geim and his team made their synthetic gecko adhesive by fabricating a **tidy** array of microscale hairs out of polyimide, a flexible and wear-resistant plastic. When <u>mounted</u> on a flexible base, the arrangement and <u>density</u> of the hairs maximize the number of hairs contacting a surface.

"The smaller the hairs are, and the more of them you have, the greater the adhesion," notes Ron Fearing, an engineer at the University of California, Berkeley.

Unlike a gecko's feet, however, the tape begins to lose its adhesive power after about five <u>applications</u>. Geim blames this <u>shortcoming</u> on polyimide's hydrophilicity, that is, its tendency to attract water. With repeated applications, some of the gecko tape's hairs get **soggy**, bunch together, and then <u>clump</u> onto the tape's base. This happens even when the tape is attached to surfaces that are dry to the touch, because they carry a layer of water two or three atoms thick.

soggy: very wet; soaked

By using hydrophilic material, Geim departed from the gecko's design—its setae are made of keratin, a so-called hydrophobic protein that repels water. Geim says hydrophobic materials, which include silicone and polyester, are more difficult to <u>mold</u> into setae-like structures than is polyimide. Even so, both he and Fearing agree, it will take water-repellant substances to produce a long-lasting gecko tape.

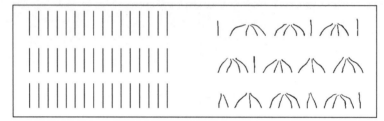

Left: straight, separate hairs on gecko tape—good adhesion
Right: clumped hairs on gecko tape—poor adhesion

 ## What's the Point?

Check your understanding of the text. Based on the reading, are the following statements true (T) or false (F)?

1. _____ Other gecko-like adhesives have been developed before.

2. _____ The newly developed "gecko tape" is three times as strong as natural gecko feet.

3. _____ Gecko tape could be used like a bandage on human skin.

4. _____ The new synthetic (artificial) tape uses van der Waals forces for adhesion.

5. _____ For repeated use, the synthetic gecko tape is more effective than real gecko feet.

6. _____ Gecko tape is made of the same chemical material as real gecko feet.

7. _____ Water creates problems for the synthetic adhesive tape.

8. _____ Gecko tape can be made from different materials with equal ease.

Understanding Words and Phrases

I. The text you just read uses the following words: hydrophilicity, hydrophilic, hydrophobic. *The text of the article explains the meanings of these words. Based on the explanations in the text, answer the following questions.*

1. What is a hydrophilic material? What is a hydrophobic material?

2. Can you give some examples of each type of material?

3. These words contain the Greek roots *hydro, phil, phob*. What is the meaning of each root?

4. Based on your knowledge of these roots, what kind of person is an *Anglophile?*

5. The word *phobia* is used to describe an abnormal psychological fear. What is *hydrophobia?*

   Are you familiar with the words *claustrophobia, agoraphobia, acrophobia, xenophobia*? If not, look up these words in the dictionary.

   Do you know anyone who has one of these phobias?

II. Words that have similar meanings are called synonyms. *Match each word or phrase in the left column with the correct synonym on the right. The words on the left are underlined in the article for easy reference. If a word has more than one meaning, use the meaning that is used in the article.*

_____	1. scurry	a.	imitate
_____	2. emulate	b.	shape
_____	3. dangle	c.	stick together
_____	4. branch	d.	co-worker
_____	5. colleague	e.	divide
_____	6. nestle	f.	amount of something in a space
_____	7. bear	g.	rush
_____	8. density	h.	hang
_____	9. clump	i.	tolerate, support
_____	10. mold	j.	snuggle

III. Words that have the opposite meanings are called antonyms. *Match each word or phrase in the left column with the correct antonym on the right. The words on the left are underlined in the article for easy reference.*

_____	1. claim	a.	simple
_____	2. intricate	b.	advantage
_____	3. application	c.	deny
_____	4. mount	d.	detach
_____	5. shortcoming	e.	removal

IV. Answer these questions or do the requested tasks. The underlined words are also underlined in the article for easy reference. Respond in full sentences.

1. New Yorkers often <u>scurry</u> along the streets. Give another example of someone or something that scurries.

   _____

2. Whom would you like to <u>emulate</u>?

   _____

3. What did Christopher Columbus <u>claim</u> about the shape of our planet?

   _____

4. List some major <u>branches</u> of science. Does each item on your list <u>branch</u> into further divisions? Give some examples.

   _____

5. Name or describe something that you just can't <u>bear</u>.

   _____

6. A computer circuit board has <u>intricate</u> electrical connections. What other things have an intricate design or mechanism?

   _____

7. A joey (a baby kangaroo) can <u>nestle</u> comfortably in its mother's pouch. Give some examples of things that can nestle in or next to other things.

   _____

8. What usually has a higher population <u>density</u>—a city or a suburb?

   _____

9. A camera can be <u>mounted</u> on a tripod for stability. Describe something else that can be mounted?

   _____

10. What are some <u>shortcomings</u> of living in a large city? What are some advantages?

_____

11. When you cut or scrape your skin, what can you <u>apply</u> to the wound to help it feel better or heal faster?

_____

12. When it's humid, sugar can <u>clump</u>. What else can clump?

_____

13. Children like to <u>mold</u> clay into various shapes. What other things, tangible or intangible, can be molded into desired forms?

_____

14. Sometimes, a small bell <u>dangles</u> from a collar around a cat's neck. Give an example of something else that dangles.

_____

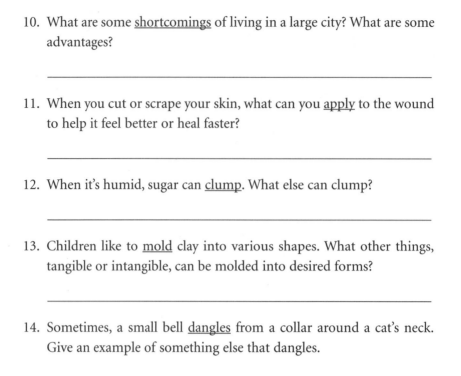

✓ Grammar Check

The Double Comparative Form

The article contains the sentence, "**The smaller** the hairs are, and **the more** of them you have, **the greater** the adhesion." What does this sentence mean?

The words in bold type are the *double comparative* form. This form consists of two parts: Both parts are comparative forms of an adjective or adverb, and both parts begin with the definite article *the*. The first part presents a condition, and the second part tells the result of that condition. So, the sentence in the article means the following: If the hairs become smaller and smaller, then the adhesion becomes greater and greater; and, if there are more and more hairs, then the adhesion becomes greater and greater.

Here are some well-known expressions:

"Yes, bring your friend to the New Year party. **The more, the merrier!**"

"Get medical attention as soon as you can. **The sooner, the better.**"

"**The more** things change, **the more** they stay the same." (Marcel Proust)

Note: if the result is in the future, the correct tense of the condition is simple present, the same as with *if* expressions. For example: *The faster you walk, the sooner you will get there. If you walk faster, you will get there sooner.*

Complete the following sentences with your own words.

1. The more I read in English, _____.

2. The closer a planet is to the sun, _____.

3. The heavier something is, _____.

4. The more energy we use, _____.

5. The larger the world's population gets, _____.

6. The more we learn in science, _____.

7. _____, the sooner I will get it done.

8. _____, the better we will get along.

9. _____, the higher the chance for success.

10. _____, the more complex it becomes.

11. _____, the faster it can go.

12. _____, the more well-behaved they are.

Let's Talk about It

1. Why do you think it took scientists so long to pin down the secret of geckos' ability to stick to surfaces? What modern tools have helped scientists figure out this puzzle?

2. What are the unique characteristics of geckos' sticking power?

3. What are some of the difficulties of manufacturing an artificial gecko-type adhesive?

4. Why are geckos the envy of rock climbers?

5. Who else might envy geckos' special ability? Why? What could it be used for?

What Do You Think?

Can you think of some people who would benefit from a gecko-type adhesive? Imagine that you are a scientist or engineer working on developing such an adhesive. You need money to fund your research. Write a proposal to an organization, company, or government agency that might be interested in funding your work. Persuade this organization to fund your work. Convince them that your work is important, that it is achievable, and that it will greatly benefit this organization.

Expansion Activities

1. What does a gecko look like? Is there more than one species of gecko? What do their feet look like? To get answers to these questions, visit the Internet and try to find pictures of geckos and their feet. Most school, university, and public libraries have free Internet access. Or, see if you can find pictures of geckos in an encyclopedia.

2. Would you like to learn more about gecko-inspired adhesive? Look for information about it on the Internet. See if you can find answers to the following questions:

   What is the current status of this research?
   What are some possible applications of this adhesive?

   Report your findings to the whole class in the form of a brief oral presentation. You can do this either alone or with a partner.

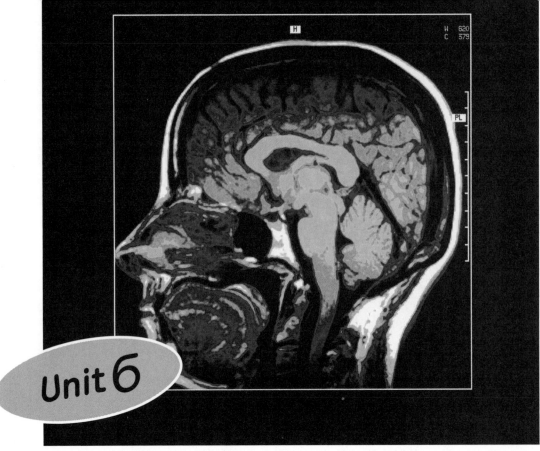

Brain power.

Time and the Brain

Search Your Knowledge

1. Have you ever had the feeling that time goes by more and more quickly every year? Do you know other people who feel the same way?

2. Why do you think this happens? Are we just imagining this feeling or is it real?

3. Can you estimate how much time goes by without looking at a clock? How well can you do that?

The text that follows reports on scientific research into how our brain measures time and how our sense of time may change as we age. The article was published in the *New York Times*. Can you understand the underlined words in the text? These words may be new to you; therefore, they are underlined so that you can find them easily later on, if you wish to refer to them again. See if you can figure out what they mean from context or from the other words and meanings around the underlined word. The words are also included in the vocabulary exercises under Key Words and Understanding Words and Phrases.

The boldfaced words in the text are glossed in the margin. These non–high frequency vocabulary words or phrases are helpful to understanding the reading.

This article is quite challenging. It is difficult even for native English speakers because it describes new scientific concepts and uses technical terms (for instance, *midbrain, basal ganglia, higher cortex, higher cortical area,* and *frontal lobes* are different parts of the brain). But you can still enjoy the overall message of this story. Don't worry about the technical details; just try to understand the main ideas. The title and the illustration, a chart, will help you become familiar with the context. Skim the article first before reading it. (For tips about skimming, see How to Read on page xiii.)

KEY WORDS

Try to figure out the meanings of the underlined words from the following examples.

   1. a. The <u>interval</u> between Christmas and New Year's Day is one week.
       b. The apple trees in the orchard are planted at equal <u>intervals</u>.
           What is an *interval?*

   2. a. People from different cultures <u>perceive</u> the world around them
           in different ways. They have different <u>perceptions</u> about what is
           right or wrong, beautiful or ugly, important or not important.
       b. Dogs can <u>perceive</u> some sounds that people can't hear.
           Can you think of other words with the same meanings as *perception* or *perceive?*

Running Late? Researchers Blame Aging Brain

Sandra Blakeslee

from the *New York Times*

Can March be almost over? Didn't they play the Super Bowl just last week? It sometimes seems that with each passing year, the days and weeks **zip by** more quickly.

zip by (idiom): go by very fast

If you have ever had this feeling, you are not imagining it. Studies of human time perception show that age-related changes in the nervous system alter one's sense of time; it really does seem to move more quickly with age. At a meeting of the Society for Neuroscience in New Orleans in November, a psychologist, Dr. Peter A. Mangan, reported on a study in which he asked people in different age groups to estimate when three minutes had passed by silently counting one-one-thousand, two-one-thousand, three-one-thousand and so on. People in their early 20s were accurate within three seconds, and some got it exactly right. People in their 60s estimated that three minutes **were up** after 3 minutes and 40 seconds had passed. Middle-aged subjects fell in between but, like the older people, all underestimated the passage of time.

were up (idiom): were over; ended

This phenomenon has led some researchers to suspect that the brain contains a special clock that tracks time intervals in the range of seconds to minutes. A Duke University neuroscientist, Dr. Warren Meck, and a graduate student, Matthew Matell, have now proposed a model of this clock based on studies of human brain anatomy.

According to their theory, a cluster of neurons in the midbrain collects time signals from all over the brain and coordinates those that occur at the same time and involve singular events or perceptions. The neurons also establish the start and finish of various time intervals that the brain is interested in measuring, such as how long it should take before a red traffic light turns green. Moreover, a brain chemical called dopamine regulates this clock. Add dopamine and the clock runs faster; take it away, and the clock slows down.

Defects in this clock could help explain human ailments like dyslexia, hyperactivity, Parkinson's disease and schizophrenia. It could explain why, in an automobile accident, three seconds can feel like three minutes, why old people in nursing homes are often confused about time, and even how some drugs like cocaine and amphetamine give the sense of "speed" while others, including marijuana, subjectively slow down the passage of time.

HOW TIME FLIES

Researchers set out to test the popular hypothesis that time seems to go faster as people age and when they are busy. They asked three groups of people, divided by age, to estimate when three minutes were up. In one part, the subjects counted the seconds; in the other, they stayed busy performing a task similar to sorting mail. Here are the average times that actually elapsed.

Counting the time

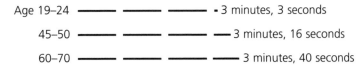

Age 19–24 ——— ——— ——— • 3 minutes, 3 seconds

45–50 ——— ——— ——— — 3 minutes, 16 seconds

60–70 ——— ——— ——— — 3 minutes, 40 seconds

While performing a task

Age 19–24 ——— ——— ——— —— 3 minutes, 46 seconds

45–50 ——— ——— ——— —— • 4 minutes, 3 seconds

60–70 ——— ——— ——— —— — 4 minutes, 46 seconds

```
      1      2      3      4      5
            minutes
```

Source: Dr. Peter A. Mangan

An American psychologist, Dr. Hudson Hoagland, first suspected the existence of an interval clock in the 1930s, when his wife ran a high fever. Mrs. Hoagland complained that her husband had been out of the room for a very long time when he had actually been gone for only a few moments. Curious, Dr. Hoagland asked his wife to estimate when a minute has passed. After 37 seconds, she said that the time was up. And as her temperature rose, she counted faster.

In further experiments, Dr. Hoagland found that he could <u>retard</u> an individual's sense of time by 20 percent by applying heat to the person's brain.

Other researchers later found that lowering a person's body temperature by two or three degrees could speed up the subjective sense of time.

The idea that there is a clock measuring <u>intervals</u> in the range of seconds to minutes (in addition to the circuits that measure tenths of hundredths of seconds and the circadian clock) makes a lot of sense. The ability to estimate short durations of time is critical for learning and survival, said Dr. John Gibbon of the New York State Psychiatric Institute and Columbia University.

People unconsciously monitor the timing of external events and respond to them. For example, Dr. Meck said, "suppose you are sitting at a red light, waiting for it to turn green," adding: "At a certain point, based on past experience, you will begin to put your foot on the gas in <u>anticipation</u> that the light is about to turn. Unconsciously, you are counting the seconds, without looking at your watch. But if the light fails to turn green in the expected amount of time, you start **fretting,** wondering if it is working properly. If enough time passes, you may decide to run the red light."

fretting: being uneasy or upset; worrying

People use interval clocks when engaged in music or sports. Basketball players, Dr. Meck points out, keep track of time in their brains, knowing that they will be penalized under certain circumstances if they hold the ball for longer than several seconds without dribbling or passing. Musicians simultaneously measure not just the beat but the phrase, the crescendos and innuendoes. When jazz players shade the time in violation of strict beats, it makes the music interesting, Dr. Gibbon said.

In the case of the basketball player, different parts of the brain are working on different tasks. Cells in the visual system are controlling movements. Cells in the auditory system are listening for information from teammates on what to do next. Each of these specialized cell circuits carrying out different jobs tends to oscillate or fire at different rates. Some might be firing 5 times a second, others up to 40 times a second. It is as if they are operating independently on different time scales yet the basketball player's brain must <u>integrate</u> them so that he or she can decide what to do with the ball.

For this task of <u>coordination</u>, Dr. Meck and his assistant have nominated a structure in the midbrain called the striatum, which is loaded with spiny neurons, so called because their projections are thick with spines. Such neurons are well connected in that each one—and there are thousands of them—is linked to tens of thousands of other cells via dendrites coming from other parts of the brain. The dendrites are the slender spines that help brain cells communicate. They detect oscillations for cell firing rates that occur all over the brain, Dr. Meck said, "and the question has been what the heck do they do with them?"

"We think spiny neurons <u>integrate</u> these signals," Dr. Meck said, and, based on previous experience of what is important, select those that are beating at the same frequency and synchronize them. This collective timing signal is sent to higher cortical areas where, in a grand **loop** from the brain's basal ganglia to its frontal lobes, <u>perceptions</u> and actions are

loop: a closed circular path

<u>coordinated</u> and acted upon. When a person is performing several tasks at once and needs to measure time, spiny neurons parcel out the tasks, Dr. Meck said.

The key to how this clock works—or fails to work—is dopamine. When the brain notices something new or rewarding, dopamine made in a nearby region called the substantia nigra is released into the spiny neurons, which become excited and begin to integrate time signals. In this way, the brain learns to anticipate events seconds or minutes into the future.

Animal and human experiments support the existence of the short-interval circuit, Dr. Meck said. For example, rats trained to press a lever at regular intervals to get food lose the ability when their dopamine-producing cells are removed. When the rats are given a synthetic form of dopamine, the ability is restored. In brain imaging experiments by Dr. Sean Hinton at Duke, people were asked to estimate when 11 seconds were up and to squeeze a ball just before and after this interval. The loop from the midbrain, where the source of dopamine and spiny neurons reside, to the higher cortex was activated each time they estimated 11 seconds had gone by.

The Days Grow Short as We Near December

The interval clock has drawn the interest of medical researchers. Dr. Guinevere Eden of Georgetown University Medical Center in Washington said that dyslexia was basically a timing problem throughout the brain and that for dyslexics difficulty in learning to read was just one manifestation of a more widespread defect. Some dyslexics have a problem with time, she said. They come late to appointments and have trouble keeping rapidly moving events in proper chronological order.

A study in the Feb. 7 issue of the British medical journal *The Lancet* found that people with attention deficit hyperactivity disorder tend to have smaller than normal frontal lobes and basal ganglia, the loop that Dr. Meck and his colleagues think is the interval time keeper.

People with Parkinson's disease lose cells that make dopamine and their interval clocks are **thrown off,** Dr. Meck said. They have tremors, difficulty in starting movements, rigid muscles and problems perceiving time accurately, all of which can be reversed with drugs that supply dopamine to the brain.

thrown off: upset or confused, usually due to something unexpected

Researchers agree that drugs affect the brain's sense of time. Cocaine and methamphetamine both increase the amount of dopamine and speed up the interval clock or its equivalent elsewhere in the brain. A similar chemical cascade may happen during an accident, when dopamine and other neurotransmitters flood the brain, Dr. Mangan said. Time seems to stand still or move incredibly slowly. Conversely, drugs that reduce the amount of dopamine in the brain, like Haloperidol, and Clozapine, used to treat schizophrenia, and marijuana, slow the interval clock, Dr. Meck said.

It is in growing older that most people will experience the vagaries of the interval clock. Studies show that dopamine levels gradually begin to fall when people are in their 20s and decline through old age, said Dr. Mangan, a psychologist at Clinch Valley College of the University of Virginia in Wise, VA. As shown by the experiments reported by Dr. Mangan in New Orleans, the older you get the more you are going to feel that time is passing by quickly. The good news is that spring will come sooner and sooner.

Source: Sandra Blakeslee, "Running Late? Researchers Blame Aging Brain," *New York Times*, March 24, 1998. Copyright © 1998 New York Times.

 ## What's the Point?

I. Check your understanding of the text. Based on the reading, are the statements true (T) or false (F)?

1. _____ People who are 35–45 years old estimate the passage of time accurately.

2. _____ Some drugs can affect the brain clock and change the way people perceive time.

3. _____ Dr. Meck and his student at Duke University were the first to propose the idea of a biological interval clock in the brain.

4. _____ High body temperature can make it seem like more time goes by than the actual amount of time.

5. _____ Without looking at a watch, we can tell if a movie lasts three hours instead of two hours with the help of the brain's interval clock.

6. _____ Some scientists believe that a collection of nerve cells in the spine organizes and manages many time signals from different parts of the brain.

7. _____ Being able to estimate short intervals of time is important only to athletes.

8. _____ Some serious diseases are caused by chemical imbalances in the brain's interval clock.

9. _____ Some medical disorders may be due to the interval clock's abnormal size.

10. _____ As we get older, our brain produces more dopamine.

II. Check your understanding of details. Complete the sentences on the left by choosing the best phrase on the right.

_____ 1. There is a region in the brain that measures

a. various activities.

_____ 2. The idea that a short-interval "clock" may exist in the brain is supported by

b. an increase in the amount of dopamine.

_____ 3. The brain's interval clock is made up of

c. short time intervals.

_____ 4. The interval clock is regulated by

d. the chemical dopamine.

_____ 5. People use the interval clock to perform

e. experiments with people and animals.

_____ 6. An accident may cause

f. a collection of neurons.

💡 Understanding Words and Phrases

I. Think of a few examples for each of the following items. Key words are underlined. They are also underlined in the article for your reference.

1. Something that <u>retards</u> the growth of a plant: _____.

2. Things that come in a <u>cluster</u>: _____.

3. Things that can have <u>defects</u>: _____.

4. Things that can be <u>integrated</u> together (or with other things): _____.

5. Something you <u>anticipate</u>: _____.

6. Events that can occur in <u>chronological</u> order: _____.

7. Things that you <u>coordinate</u>: _____.

8. <u>Ailments</u> that older people have: _____.

II. These two exercises deal with figurative expressions. Answer the questions in each exercise.

1. Here is a phrase from the article: *"The days grow short as we near December."*
   This phrase has two meanings, one *literal* and one *figurative*.
   What is the literal meaning?
   Try to figure out the figurative meaning by answering these questions:
      What happens to the length of daylight as we get closer to the month of December?
      What happens to the length of our days as we age?
      What are the similarities between "nearing December" and "growing older"?

2. The title of the chart in the article is *"How Time Flies."*
   There is a saying in English: *"Time flies when you're having fun."*
   What do you think it means? Do you agree that time passes very quickly when you are having a good time?

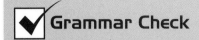

Grammar Check

Gerunds

Look at the following examples from the article. Why is the *-ing* form used in each case (the underlined words provide a hint)?

Type 1:
lowering . . . body temperature . . . could speed up the subjective sense of time

Type 2:
you <u>start</u> **fretting**

Type 3:
to estimate . . . <u>by</u> **counting**
interested <u>in</u> **measuring**
similar <u>to</u> **sorting**
the ability . . . is critical <u>for</u> **learning**
hold the ball . . . <u>without</u> **dribbling** or **passing**

Type 4:
they <u>stayed busy</u> **performing** a task
you are <u>sitting</u> at a red light **waiting** for it to turn green
Dr. Meck <u>said</u> . . . **adding** . . .
they . . . <u>have trouble</u> **keeping**
they <u>have . . . difficulty (in)</u> **starting**
they <u>have . . . problems</u> **perceiving**

The words in bold are called **gerunds**. A **gerund** is an *-ing* form of a verb that is used as a noun; for example, *Reading is fun.* (You are familiar with several other *-ing* forms of verbs. One form is a *present participle* used as a verb in the present progressive tense; for example, *He is **watching** a movie.* Another form is a present participle used as an adjective; for example, *I heard some **interesting** news.* Yet another form is used in adverbial phrases; see Grammar Check: Adverbial Phrases in Unit 4.)

When English speakers want to use a verb-like noun, they often have to choose between a gerund (like *reading*) and an infinitive (like *to read*). In this unit, we'll review the use of gerunds.

Following is a summary of the common uses of a gerund, shown in the preceding text examples.

1. **Type 1, subject of a sentence.** Either a gerund or an infinitive can be the subject of a sentence, but gerunds are much more common as subjects than infinitives. For example:

   Studying a foreign language is required in this school.

   Worrying about it won't help you.

2. **Type 2, after certain verbs.** Some verbs (such as *enjoy*) must be followed by gerunds, while others (such as *want*) must be followed by infinitives; still others (such as *start*) may be followed by either a gerund or an infinitive. There is no rule for distinguishing such verbs; they must be learned and memorized like vocabulary. For example:

   Janet <u>enjoys</u> ***traveling*** around the world.

   Bill doesn't <u>mind</u> ***doing*** the dishes after dinner.

   Sometimes, he <u>likes</u> ***being*** alone. Or: Sometimes, he <u>likes</u> ***to be*** alone.

3. **Type 3, after prepositions.** Fortunately, this is an easy rule, without exceptions. For example:

   I plan <u>on</u> ***taking*** vacation next fall.

   Let's take a walk <u>after</u> ***eating*** this large meal.

4. **Type 4, after special expressions.** There are several kinds of such expressions.

   a. One kind expresses the ease or difficulty of doing something: *have fun, have a good time, have a bad time, have trouble, have a difficult time,* etc. For example:

> I always <u>have a great time</u> ***talking*** to you!
>
> John <u>had a hard time</u> ***moving*** to a new city.

   b. Another kind describes two simultaneous actions, often with an expression of time or place, where the gerund shows the main activity: *spend* or *waste* (time or money) doing something; *sit, stand,* or *walk* (somewhere) doing something. For example:

> Don't <u>waste</u> money ***fixing*** this old TV set; buy a new one.
>
> We <u>walked</u> along the street ***taking*** in the new sights.

   c. A third kind gives the idea of discovering someone doing something, often something undesirable: *discover, catch, find* (someone or something). For example:

> Little Susie was <u>caught</u> ***eating*** a cookie before dinner.

   d. Another well-known special expression has the form *go + gerund* and is used for leisure or sport activities: *go shopping, go hiking.* For example:

> Do you want to <u>go</u> ***skiing*** in Colorado this winter?

Exercise 1: Gerunds as Subjects

It is common to see an infinitive phrase used with the word it *as the subject of a sentence:* It *is easy* to find my house. *In such sentences,* it *refers to the infinitive phrase. Although an infinitive can be used as the subject of a sentence—*To find *my house is easy—it is more common to see a gerund as the sentence subject:* Finding *my house is easy. For each item, change the given sentence with an infinitive phrase to a sentence with a gerund phrase as the subject.*

1. It's good for you to exercise regularly.

2. It is dangerous to climb mountains, especially in the winter.

3. It takes a long time to learn to ride a bicycle. Once you learn, it's easy to ride it.

4. Is it impolite to ask a lot of personal questions if you don't know the person very well?

5. It's not as easy as it looks to figure-skate beautifully.

6. Is it more difficult for an adult to learn a foreign language than for a child?

Now, complete each following sentence with an infinitive phrase, using your own words. Then, rewrite the sentence using a gerund as the subject.

7. It is very important _____

_____.

8. Is it a good idea _____

_____?

9. It is usually wrong _____

_____.

10. It sounds like a lot of fun _____

    _____ .

11. It's impossible (for me) _____

    _____ .

12. It is better _____ than _____

    _____ .

Exercise 2: Gerunds Following Verbs, Prepositions, and Special Expressions

Complete these sentences with either a gerund or an infinitive form of the given verb. Why did you choose a particular form?

1. The boy is refusing *(get on)* _____ the elevator because he is claustrophobic.

2. He was telling her about his childhood, *(describe)* _____ everything in colorful detail.

3. Do you want *(go + bowl)* _____ after work this evening?

4. Melanie prefers *(hike)* _____ to *(swim)* _____ .

5. Pamela's brother advised her *(see)* _____ a doctor about her eye problem. She promised *(see)* _____ an ophthalmologist next week.

6. Greg can put a great deal of effort into *(solve)* _____ difficult physics problems.

7. Mr. and Mrs. Wright often discuss *(stay)* _____ at home for their vacation, but they usually end up *(travel)* _____ to interesting and exotic places.

8. The young campers had fun *(swim)* _____ in the lake and *(learn)* _____ *(catch)* _____ fish.

9. Because it's raining, I suggest *(go + shop)* _____ at a mall instead of *(go + sightsee)* _____.

10. Thank you for *(take care of)* _____ my house and yard while I was on vacation.

11. The Martins are considering *(enroll)* _____ in a course about *(invest)* _____ money.

12. When we came home late at night, we found a stray cat *(sleep)* _____ on the front porch.

13. My professor doesn't tolerate *(laugh)* _____ at other students' mistakes.

14. She has a hard time *(drive)* _____ at night because she has difficulty *(see)* _____ in the dark.

15. The family finally succeeded in *(reach)* _____ their destination by *(follow)* _____ the map and *(ask)* _____ local residents for directions.

16. Oscar loves *(spend)* _____ his free time *(watch)* _____ movies and *(play)* _____ computer games.

Exercise 3

Complete these sentences in your own words. Provide your own verb in the correct form—gerund or infinitive.

1. You can attract birds to your garden by _____.

2. We tried _____.

3. Mr. and Mrs. Lasky plan _____.

4. Carol sat in the rocking chair _____.

5. Please close the windows before _____.

6. We remained in class _____.

7. I really miss _____.

8. My wife can't stand _____.

9. I'm really looking forward to _____.

10. The students in the class had a problem _____.

Exercise 4

Create your own sentences from the given combinations of verbs and expressions. You may use any noun, pronoun, or person's name; you may also use any verb tense or modal. The first item is done for you as an example.

1. begin + operate

   At 8:30 A.M., the surgeon <u>began</u> **operating** on the vice-president's heart.

   Or: At 8:30 A.M., the surgeon <u>began</u> **to operate** on the vice-president's heart.

2. stand (somewhere) + watch

3. decide between + (any two actions)

4. spend an hour + explain

5. avoid + go + dance

6. have trouble + understand

7. catch (someone or something) + hide

8. hate + go + camp

9. delay + do laundry

10. continue + play

Let's Talk about It

1. Take a look at the chart presented in the article. Researchers asked people to estimate the passage of time under two different conditions. What were the two conditions?

   Under what conditions are people more accurate when they estimate time?

   Do these conditions affect young and old people in the same way or in different ways?

   Based on what you read and on your own experience, why do you think our time estimates are different under the two conditions above?

2. Have you noticed that time goes faster when you are busy? How about when you are happy or sad?

   Give some examples from your experience. Why do you think there is a difference, if any?

3. Does it make sense to you that time passes more quickly as we age because of changes in the "interval clock" in the brain? Why or why not?

What Do You Think?

As you read in this article, scientists believe that our perception of time changes as we get older because of biological changes in the brain. Do you think there might be other reasons why time seems to move faster as we age? Write an essay arguing your case.

Expansion Activities

The text described an experiment where researchers measured how well people estimate time. See if you can replicate this experiment. Ask two (or more) friends or relatives of different ages to estimate when three minutes go by. Report your findings in class.

Highway bridges in Salt Lake City, Utah.

Engineering Achievements

Search Your Knowledge

1. What was life like 100 years ago? How much of your life today is affected by engineering accomplishments?

2. What do engineers do?

3. What are some of the engineering achievements of the 20th century? List as many as you can.

4. Think about what life might be like without these developments. Which engineering achievements are the most important? Why?

5. <u>Class activity:</u> Rank the engineering achievements in order of importance. Discuss the choices and agree or vote on the order.

Now, compare your class results to the list of engineering achievements that was put together by the National Academy of Engineering in the United States. Read the document on pages 106–10 from the Academy, which selected the top engineering feats of the 20th century. The document was issued as a *press release* to various news services and press clubs, including the National Press Club in Washington, DC. Can you understand the underlined words in the text? These words may be new to you; therefore, they are underlined so that you can find them easily later on, if you wish to refer to them again. See if you can figure out what they mean from context or from the other words and meanings around the underlined word. The words are also included in the vocabulary exercises under Key Words and Understanding Words and Phrases.

The boldfaced words in the text are glossed in the margin. These non–high frequency vocabulary words or phrases are helpful to understanding the reading. Read the entire text before doing the exercises.

NATIONAL ACADEMY OF ENGINEERING

2101 Constitution Avenue, NW
Washington, DC 20418
http://www.national-academies.org

Date: February 22, 2000
Contacts: Charles E. Blue, American Association of Engineering Societies
             999-555-2237 X16; e-mail *xxxx@aaes.org;*
             cell phone 999-555-8865 x55555
             Donald Lehr, National Engineers Week
             999-555-8200; e-mail *xxxx@compuserve.com;*
             cell phone 999-555-4058
             Robin Gibbin, National Academy of Engineering
             999-555-1562; e-mail *xxxx@nae.edu*

[EMBARGOED: NOT FOR PUBLIC RELEASE BEFORE 1:00 P.M. EST TUESDAY, FEB. 22]

National Academy of Engineering Reveals Top Engineering Impacts of the 20th Century: Electrification Cited as Most Important

WASHINGTON—One hundred years ago, life was a constant <u>struggle</u> against disease, pollution, deforestation, treacherous working conditions, and <u>enormous</u> cultural divides unbreachable with current communications technologies. By the end of the 20th century, the world had become a healthier, safer, and more productive place, primarily because of engineering achievements.

Speaking **on behalf of** the National Academy of Engineering (NAE), astronaut/engineer Neil Armstrong today announced the 20 engineering achievements that have had the greatest impact on quality of life in the 20th century. The announcement was made during National Engineers Week 2000[1] at a National Press Club luncheon.

on behalf of:
speaking for

The achievements—nominated by 29 professional engineering societies—were selected and ranked by a distinguished panel of the nation's top engi-

---

[1] National Engineers Week 2000 is held February 20–26. National Engineers Week always falls around the birth date of George Washington, who was a surveyor and is often cited as America's first engineer.

neers. Convened by the NAE, this committee—chaired by H. Guyford Stever, former director of the National Science Foundation (1972–76) and Science Advisor to the President (1973–76)—worked in anonymity to ensure the <u>unbiased</u> nature of its <u>deliberations</u>.

"As we look at engineering breakthroughs selected by the National Academy of Engineering, we can see that if any one of them were removed, our world would be a very different—and much less <u>hospitable</u>—place," said Armstrong. Armstrong's announcement of the top 20 list, which includes space exploration as the 12th most important achievement, covers an incredibly broad spectrum of human endeavor—from the vast networks of electrification in the world (No. 1), to the development of high-performance materials (No. 20) such as steel alloys, polymers, synthetic fibers, composites and ceramics. In between are advancements that have revolutionized the way people live (safer water supply and treatment, No. 4, and health technologies, No. 16); work (computers, No. 8, and telephones, No. 9); play (radio and television, No. 6); and travel (automobile, No. 2, airplane, No. 3, and interstate highways, No. 11).

In his statement delivered to the National Press Club, Armstrong said that he was <u>delighted</u> to announce the list of the greatest achievements to underscore his commitment to advancing the understanding of the critical importance of engineering. "Almost every part of our lives underwent <u>profound</u> changes during the past 100 years thanks to the efforts of engineers, changes impossible to imagine a century ago. People living in the early 1900s would be amazed at the advancements wrought by engineers," he said, adding, "as someone who has experienced **firsthand** one of engineering's most incredible advancements—space exploration—I have no doubt that the next 100 years will be even more amazing."

> **firsthand:** directly, personally

The NAE notes that the top achievement, electrification, powers almost every <u>pursuit</u> and enterprise in modern society. It has literally lighted the world and impacted countless areas of daily life, including food production and processing, air conditioning and heating, refrigeration, entertainment, transportation, communication, health care, and computers.

Many of the top 20 achievements, given the immediacy of their impact on the public, seem obvious choices, such as automobiles, at No. 2, and the airplane, at No. 3. These achievements, along with space exploration, the nation's interstate highway system at No. 11, and petroleum and gas technologies at No. 17, made travel and mobility-related achievements the single largest segment of engineering to be recognized.

Other achievements are less obvious, but nonetheless introduced changes of staggering proportions. The No. 4 achievement, for example, the availability of safe and <u>abundant</u> water, literally changed the way Americans lived and died during the last century. In the early 1900s, waterborne diseases like typhoid fever and cholera killed tens-of-thousands of people annually, and dysentery and diarrhea, the most common waterborne disease, were the third largest cause of death. By the 1940s, however, water treatment and distribution systems devised by engineers had almost totally eliminated these diseases in America and other developed nations. They also brought water to vast tracts of land that would otherwise have been uninhabitable.

No. 10, air conditioning and refrigeration technologies, underscores how seemingly commonplace technologies can have a staggering impact on the economy of cities and worker productivity. Air conditioning and refrigeration allowed people to live and work effectively in <u>sweltering</u> climates, had a <u>profound</u> impact on the distribution and preservation of our food supply, and provided stable environments for the sensitive components that <u>underlie</u> today's information-technology economy.

Referring to achievements that may escape notice by most of the general public, Wm. A. Wulf, president of the National Academy of Engineering, said, "Engineering is all around us, so people often take it for granted, like air and water. Ask yourself, what do I touch that is not engineered? Engineering develops and delivers consumer goods, builds the networks of highways, air and rail travel, and the Internet, mass produces antibiotics, creates artificial heart valves, builds lasers, and offers such wonders as imaging technology and conveniences like microwave ovens and compact discs. In short, engineers make our quality of life possible."

Selection Process

The process for choosing the greatest achievements began in the fall of 1999, when the National Academy of Engineering, an <u>enormous</u> non-profit organization of outstanding engineers founded under the congressional charter that established the National Academy of Sciences, invited discipline-specific professional engineering societies to nominate up to ten achievements. A list of 105 selections was given to a committee of academy members representing the various disciplines. The panel convened on December 9 and 10, 1999, and selected and ranked the top 20 achievements. The overarching criterion used was that those advancements had made the greatest contribution to the quality of life in the past 100 years. Even though some of the achievements, such as the telephone and the automobile, were invented in the 1800s, they were included because their impact on society was felt in the 20th century.

The Achievements

Here is the complete list of achievements as announced today by Mr. Armstrong:

1) **Electrification**—the vast networks of electricity that power the developed world.

2) **Automobile**—revolutionary manufacturing practices made the automobile the world's major mode of transportation by making cars more reliable and affordable to the masses.

3) **Airplane**—flying made the world accessible, spurring globalization on a grand scale.

4) **Safe and Abundant Water**—preventing the spread of disease, increasing life expectancy.

5) **Electronics**—vacuum tubes and, later, transistors that <u>underlie</u> nearly all modern life.

6) **Radio and Television**—dramatically changed the way the world received information and entertainment.

7) **Agricultural Mechanization**—leading to a vastly larger, safer, less costly food supply.

8) **Computers**—the heart of the numerous operations and systems that impact our lives.

9) **Telephone**—changing the way the world communicates personally and in business.

10) **Air Conditioning and Refrigeration**—beyond convenience, it extends the shelf life of food and medicines, protects electronics and plays an important role in health care delivery.

11) **Interstate Highways**—44,000 miles of U.S. highways allowing goods distribution and personal access.

12) **Space Exploration**—going to outer space vastly expanded humanity's horizons and introduced 60,000 new products on Earth.

13) **Internet**—a global communications and information system of unparalleled access.

14) **Imaging Technologies**—revolutionized medical diagnostics.

15) **Household Appliances**—eliminated strenuous, laborious tasks, especially for women.

16) **Health Technologies**—mass production of antibiotics and artificial implants led to vast health improvements.

17) **Petroleum and Gas Technologies**—the fuels that energized the 20th century.

18) **Laser and Fiber Optics**—applications are wide and varied, including almost simultaneous worldwide communications, non-invasive surgery, and point-of-sale scanners.

19) **Nuclear Technologies**—from splitting the atom, we gained a new source of electric power.

20) **High-Performance Materials**—higher quality, lighter, stronger, and more adaptable.

# # #

Editor's Notes:

Additional information and visuals are available at *http://www. greatachievements.org.*

Greatest Engineering Achievements of the 20th Century is a collaborative project led by the National Academy of Engineering, with the American Association of Engineering Societies, National Engineers Week, and 29 engineering societies.

The National Academy of Engineering was established in 1964, under the charter of the National Academy of Sciences, as a parallel organization of outstanding engineers. It is <u>autonomous</u> in its administration and in the selection of its members, sharing with the National Academy of Sciences the responsibility for advising the federal government. The National Academy of Engineering also sponsors engineering programs aimed at meeting national needs, encourages education and research, and recognizes the superior achievements of engineers.

Since its founding in 1951 by the National Society of Professional Engineers, National Engineers Week, a consortium of more than 100 engineering, scientific, educational societies, and major corporations, has helped increase public awareness and appreciation of technology and the engineering profession. National Engineers Week 2000 co-chairs are the American Consulting Engineers Council (ACEC), a national organization of private engineering firms, and CH2M HILL, a global engineering company specializing in water and wastewater, environmental management, transportation, telecommunications, industrial facilities, and related infrastructure.

American Association of Engineering Societies is a federation of engineering societies dedicated to advancing the knowledge, understanding, and practice of engineering, whose membership represents more than one million engineers in the United States.

What's the Point?

I. Demonstrate your understanding of some of the details in the reading. Based on the text you just read, give short answers to the questions that follow.

1. Who is Neil Armstrong? What is his profession?

2. When is National Engineers Week held? When was it first established?

3. Why was electrification chosen as the most important achievement?

4. How many engineering achievements were originally nominated for selection?

5. How many organizations participated in the project to select the top 20 achievements?

6. What is the function of the National Academy of Engineering?

II. Show your understanding of the reading. Based on the text you just read, choose the best answer to complete the statements.

1. Many people consider _____ to be the first American engineer.
   a. George Washington
   b. Neil Armstrong
   c. Wm. A. Wulf
   d. Thomas Edison

2. The area of engineering that received the most recognition on the list of top achievements is _____.

    a. power production

    b. transportation

    c. medicine

    d. agriculture

3. The most important achievement affecting public health was _____.

    a. imaging technology

    b. laser surgery

    c. antibiotics

    d. safe water supply

4. Air-conditioning and refrigeration show how something that seems ordinary can _____.

    a. be very significant

    b. be highly complex

    c. be very popular

    d. be taken for granted

5. The National Academy of Engineering is administered by _____.

    a. the federal government

    b. National Engineering Week

    c. its own members

    d. a major corporation

Understanding Words and Phrases

The words or phrases that fit these definitions are underlined in the text you just read. Use the context to try to figure out their meanings, and match them with the given definitions. Write the words or phrases in the spaces provided so that each letter fits in a separate space. Notice that each answer has a numbered letter. When you have written in all the answers, arrange the numbered letters in numerical order at the bottom of the page to spell a word that you know well!

Example: in large quantity, plentiful: a b u n d a n t

1. to be the basis (of something): __ __ __ __ __ __ __ __
                                           1

2. very large: __ __ __ __ __ __ __ __
                     2

3. a difficult battle, a big effort: __ __ __ __ __ __ __ __
                                             3

4. generous to guests; favorable to living: __ __ __ __ __ __ __ __ __ __
                                                      4

5. deep, thorough, far-reaching: __ __ __ __ __ __ __ __
                                                5

6. formal discussion and debate of an issue: __ __ __ __ __ __ __ __ __ __ __ __
                                                          6

7. greatly pleased, very glad: __ __ __ __ __ __ __ __ __
                                                 7

8. a chase, a search; a work activity: __ __ __ __ __ __ __
                                          8

9. neutral, fair, not prejudiced: __ __ __ __ __ __ __ __
                                          9

10. independent, self-governing: __ __ __ __ __ __ __ __ __
                                             10

11. terribly hot and humid: __ __ __ __ __ __ __ __ __ __
                                                 11

Alexandre Gustave Eiffel designed the Eiffel Tower for the Paris Exhibition of 1889. This was his profession: __ __ __ __ __ __ __ __ __ __ __
  1 2 3 4 5 6 7 8 9 10 11

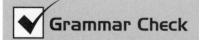

Grammar Check

Quoted Speech and Reported Speech

The press release you just read contains examples of both *quoted speech* and *reported speech*. Can you find some of these examples?

Part I: Quoted Speech

When we want to quote, or write exactly, what another person said, we write the speaker's exact words inside quotation marks ("..."). Here are some examples. In the first example, Mr. Wulf is the speaker and *said* is the reporting verb. Other reporting verbs may be used.

1. Mr. Wulf **said,** "Engineering is all around us, so people often take it for granted, like air and water."

2. The children **shouted** happily, "We want cake!"

3. "Today, the class will end early," **announced** the teacher.

4. "Are you sure?" **asked** Maria quietly.

5. "My son has been sick for five days," he **told** the doctor, "and I'm very worried about him."

Notice the basic punctuation rules for a direct quote in another sentence:

- The speaker and verb (and any adverbs) are followed by a comma before the opening quotation mark.
- The first word of the quote inside the opening quotation mark is capitalized.
- The final punctuation mark of the quote is inside the closing quotation mark.

▸ A direct quote can appear at the beginning of a sentence, and the speaker and reporting verb can be at the end. In this case, if the final punctuation mark of the quote is a period, the period becomes a comma; however, a question mark or exclamation point remain the same.

▸ The speaker and reporting verb can also appear in the middle of a direct quote. In this case, the quotation marks go around each part of the quote, the continued part of the quote is not capitalized, and the speaker and verb are surrounded by commas.

Exercise 1

Write your own sentences with direct quotes. Many different reporting verbs and adverbs provide information about the manner of the quoted speech. You have already seen some of them in the examples. Additional examples of reporting verbs and adverbs follow. Use these verbs to create <u>at least five</u> sentences; feel free to use these or other adverbs. Be sure to use correct punctuation and capitalization.

REPORTING VERB	ADVERB OR ADVERBIAL PHRASE
yelled	angrily
whispered	slowly
reported	shyly
said	to the woman
joked	to him/her

Exercise 2

Now, work with a partner to create five more sentences using different reporting verbs, such as advise, suggest, insist, tell, ask, inform, *etc.*

Part II: Reported Speech

Reported speech states the message of a speaker but does not use the speaker's exact words. Quotation marks are not used, and pronouns and verb forms may change. For example,

Quoted speech: He said, "I go shopping once a week."

Reported speech: He said that he goes shopping once a week.

In reported speech, the following rules apply:

a. The reported information is usually preceded by the word *that*, which may be omitted.

> *Example:*   She claims *(that)* she knows the answer.

b. If the reporting verb in the main sentence is in the present (or present progressive) tense, then the verb in the reported speech can be in the present tense or in the past tense.

> *Examples:*   He **says,** "I **go** to work every day."
> → He **says** that he **goes** to work every day.
>
> He **says,** "I **went** to work yesterday."
> → He **says** that he **went** to work yesterday.
>
> He **is saying,** "I **don't** work on holidays."
> → He **is saying** that he **doesn't** work on holidays.

c. If the reporting verb in the main sentence is in the past tense, then the verb in the reported speech is usually *back-shifted* from the present tense to the past tense also. This applies to perfect tenses and modals as well: present perfect is back-shifted to the past perfect, and present-form modals are back-shifted to the past form.

> *Examples:*   Armstrong **said,** "I **am** delighted."
> → Armstrong **said** that he **was** delighted.
>
> "I **have worked** all my life," she **informed** me.
> → She **informed** me that she **had worked** all her life.
>
> I **said,** "We **can't** come."
> → I **said** that we **couldn't** come.

d. In reported questions, the reported information is usually preceded by the word *whether* or *if* (informally) instead of *that*.

> *Examples:*   "Will you finish your paper on time?" my professor asked me.
> → My professor asked me **whether** I would finish my paper on time.
>
> They asked me, "Can you come to the party?"
> → They asked me **if** I could come to the party.

e. Sometimes, if the reported information represents a continued condition or a general truth, then the verb in the reported speech is not back-shifted.

Examples: Armstrong **said** that engineering **is** very important.

The teacher **told** them that the sun **rises** in the east.

Exercise 1

Part I began with five examples of quoted speech. Rewrite these five sentences as reported speech. The first sentence is done for you:

1. Mr. Wulf said that engineering is all around us, so people often take it for granted, like air and water.

2. _____ .

3. _____ .

4. _____ .

5. _____ .

Exercise 2

In Part I, Exercises 1 and 2, you created ten original sentences with quoted speech. Now restate those sentences in the form of reported speech.

 Let's Talk about It

1. On what basis, or *criterion*, were the engineering achievements selected and ranked?

2. On this basis, do you agree with the choices made by the NAE?

3. What other criteria could be used to select the top engineering achievements?

4. Based on a different criterion, how would you rank the top 20 achievements? Would you choose any other achievements that are not on the NAE list?

Activity: Work in groups of two or three students. Work with your partner(s) to select and rank the top 20 engineering accomplishments of the 20th century according to a different criterion. Think of your own, or choose one of these suggested criteria:

- ▸ technologies created in your lifetime
- ▸ the greatest impact on public safety
- ▸ the most technologically challenging
- ▸ the most original

What are your reasons for selecting the accomplishments? Present your group's results to the whole class. One student from each group can do the presentation, or each team member can present a different detail. Be prepared to explain and defend your selections.

What Do You Think?

What technological achievement or achievements have had the most profound effect on your own life? What modern technologies would you not want to live without? Why? Write an essay describing a few engineering achievements that are most important to you personally. Explain how they have affected your life.

Expansion Activities

The word *commonplace* means "ordinary, everyday." However, there is also something *extraordinary* about this word. Its letters can be combined in different ways to make many different English words—more than 80 words! See how many you can find. The words can be as long or as short as you like, but you cannot use any letters that are not in the word *commonplace*. Three of the words are given as examples.

place
common
mane

If you were standing in front of this bakery, what might you smell?

Science News and Fun

Search Your Knowledge

1. Where can you get news about science and technology?

2. Do you know of any radio or television programs that provide information about science?

3. How can you find out about such programs?

4. Have you ever looked for science-related information on the Internet?

Radio and TV programs about science and nature can be both educational and entertaining.

RADIO

Local public radio stations usually broadcast programs by National Public Radio (NPR), a publicly funded organization. NPR has many programs, including the science-related show *Talk of the Nation—Science Friday*™. The British program *BBC World Service—Science in Action* may also be heard on public radio.

TELEVISION

Local public television channels usually carry the shows *Nova, Nature,* and *National Geographic.* There are the private cable channels *Discovery Channel* and *Science Channel.*

INTERNET

Many organizations, including magazines, newspapers, and TV and radio stations, have their own Internet websites. NPR has its own website, *www.npr.org,* with a link to the *Science Friday*™ website, *www.sciencefriday.com,* where you can learn about some interesting discoveries in science. You can even listen to an entire radio show on the computer! (Some libraries provide earphones for your use.)

Get to know your local public radio and TV stations. Find out what programs they broadcast and at what time. You can call them to request a schedule of all their programs, or look them up on the Internet. Many websites offer interactive activities.

Read three summaries of stories that were discussed on the NPR radio program *Talk of the Nation—Science Friday™*. This is a two-hour show, in which each hour is devoted to a different topic. These summaries (and many more) are available on the *Science Friday™* website. Can you understand the underlined words in the text? These words may be new to you; therefore, they are underlined so that you can find them easily later on, if you wish to refer to them again. See if you can figure out what they mean from context or from the other words and meanings around the underlined word. The words are also included in the vocabulary exercises under Key Words and Understanding Words and Phrases.

The boldfaced words in the text are glossed in the margin. These non–high frequency vocabulary words or phrases are helpful to understanding the reading.

Story summary 1.

SCIENCE FRIDAY > ARCHIVES > 2000 > MARCH > MARCH 10, 2000

Smell

from *Talk of the Nation—Science Friday™*,

National Public Radio, Inc.

Take a deep breath. Smell anything? Maybe you're getting a few whiffs of early-spring air. Maybe you're dealing with the garlic-anchovy pizza on your co-worker's desk. Or maybe you're catching molecules of perfume from the woman next to you in the elevator. So how does the sense of smell work? What happens when it doesn't work right? And how are efforts to develop odor-related technology progressing?

On one level, the workings of your sniffer are pretty <u>straightforward</u>. Odor molecules have a specific shape. When they enter the nose, if those shapes match up with surface proteins on odor-sensing neurons (about 10 million of them, with over 1,000 configurations) the neuron fires. But at levels beyond that, it gets much more complicated, involving subtle mixes of chemicals, the ways that proteins <u>bind</u> together, how long they stick to receptors, genetics, and more.

"Not smelling" may seem like a joke, but 2–3 percent of the population is "anosmic"—having little or no sense of smell. About 15% of the population has some form of "odor blindness"—not being able to smell certain things that other people can smell. And since about 90% of what people perceive as "taste" is actually produced by smell (hold your nose and eat a piece of chocolate—it doesn't taste very chocolatey at all!), a lack of ability to smell can become a real quality-of-life issue.

Several companies are developing "artificial noses," sensors that can identify odor components **on the fly.** Most of these are designed for some type of quality-control <u>monitoring</u> or environmental sampling, such as making sure that fish shipments are fresh or that oil refineries aren't <u>emitting</u> compounds they shouldn't. A few inventors are working on using an artificial nose for medical diagnoses, using it to sense <u>faint</u> <u>tell-tale</u> scents given off by certain bacteria, for example. And there's even an effort afoot to create a sort of Smell-o-Vision for the Internet Age—an artificial odor generator called DigiScents.

on the fly (idiom): in a hurry

Join us on this hour of Science Friday as we find out what science knows about noses!

Source: "Smell," *Talk of the Nation—Science Friday*™, National Public Radio, Inc., March 10, 2000. *www.sciencefriday.com.* Science Friday™ © Samanna Productions, Inc.

What's the Point?

Check your understanding of the text. Based on the reading, are the statements true (T) or false (F)?

1. _____ Different odor molecules have different shapes.

2. _____ About 17–18% of the population has some problem with smell.

3. _____ People who cannot smell can still taste most food.

4. _____ Companies are developing artificial noses for people who cannot smell.

Understanding Words and Phrases

I. Based on the text you just read, choose the best definition of the underlined word or phrase. These words and phrases are also underlined in the text for your reference. If you aren't able to determine the meaning through context, use other vocabulary skills such as word parts and word form to help you with the meaning.

1. <u>straightforward</u>:	a. painful	b. unusually fast	c. simple
2. <u>bind</u>:	a. belong	b. stick	c. communicate
3. <u>faint</u>:	a. weak	b. unpleasant	c. healthy
4. <u>monitor</u>:	a. observe closely	b. escape	c. compute
5. <u>emit</u>:	a. import	b. send out	c. pollute
6. <u>tell-tale</u>:	a. interesting	b. telling a lie	c. revealing information

II. Answer the following questions about the underlined expressions that appear in the text.

1. In this story, what is a <u>sniffer</u>? Do you recognize the *doer* suffix *-er*? What does it mean <u>to sniff</u>? What animal is well-known for sniffing?

2. The story introduces and explains two new technical terms, which are unfamiliar to most English speakers. Do you know what they mean now?
   a. What does <u>anosmic</u> mean?
   b. What is <u>odor blindness</u>?

3. Based on the definition of <u>odor blindness</u>, can you guess what *color blindness* means? Do you know anyone who is color-blind?

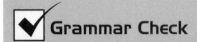

Grammar Check

Informal Register

Notice the unusual question form in the summary you just read:

> Smell anything?

What's unusual about this question form?

In English conversation, as in many languages, there are different appropriate ways of speaking in different situations—formal or informal, child or adult, professional or social, military or civilian, etc. A particular way of speaking in each situation is called a *register.* In the informal register, it is common for a speaker to speak in sentence fragments, omitting the subject or the verb, or both. The speaker can do so safely because both the speaker and listener are part of the same immediate situation; both participants can convey certain information perfectly clearly using fewer words. In the example, the radio announcer is speaking informally to his audience, and this is clear to both the radio host and the audience. Therefore, the auxiliary verb *(do)* and the subject *(you)* are omitted from the more usual, formal question form: *Do you smell anything?*

Consider this situation: A father enters the kitchen and speaks to his young son.

> *Father:*   Mmm, smells great! Mom baking a cake?
>
> *Son:*   Yeah! For my birthday!

Can you rewrite this dialogue in the grammatically correct formal register?

Exercise 1

Change the expressions in the informal register to the usual grammatical forms.

1. What's that noise? See anything outside?

2. When I count to three, start running. Ready?

3. *A:* Hey, it's almost one o'clock. You hungry?
   B: Starving! You?

4. Good luck on your test! Hope you get an A.

5. *A:* Where's Mary? She coming?
   B: Yeah, late as usual.

Exercise 2

Change the expression in the formal register to the informal register. Describe briefly a situation where such speech might occur.

1. Where's Gina? Is she not feeling well?

2. Are you tired?

3. I'll see you tomorrow. I hope your friend can come too!

4. It looks beautiful. Did you make it yourself?

5. *A:* That's a cute card. Did Maggie write it?
   B: Yes, she wrote it at school.

6. *A:* Whose turn is it? Is it mine?
   B: No, it's Pete's.

Exercise 3

Your turn! Create two expressions or short dialogs in the informal register. Describe the situations in which these conversations take place.

Let's Talk about It

1. Why do we smell some things but not others?

2. Are there animals that can smell odors that people can't?

3. Do you think that everything has a smell? What does the story suggest?

4. Is the sense of smell useful? Why or why not?

What Do You Think?

How important is the sense of smell? Can you think of one or more situations where the sense of smell was helpful to you? Write a brief story about your experiences. Tell your classmates about it.

Story summary 2.

Global Climate Change

from *Talk of the Nation—Science Friday*™,

National Public Radio, Inc.

Over 100 years ago, Swedish scientist Svante Arrhenius proposed the possibility of human-caused climate change. Since then, the world has become more and more dependent on fossil fuels for energy. Today, approximately 85% of the world's energy comes from <u>fossil</u> fuels. Meanwhile, the possibility of global long-term climate change caused by greenhouse gas emissions has moved firmly into the public consciousness.

Carbon dioxide is one of the most likely "greenhouse gases." Other gases, from methane to water vapor, can also cause the greenhouse effect. When in the atmosphere, these gases can act like a <u>thermal</u> blanket around the planet, preventing heat energy from the sun from radiating back out into space. Possible results of this warming, some scientists say, could include changing rainfall patterns worldwide, shifting agricultural zones, and melting glacier and ice caps.

In 1992, the United Nations Framework Convention on Climate Change recognized the potential problem and set an ultimate objective to stabilize greenhouse gas levels below dangerous levels (though the document didn't say what "dangerous" levels were). The Kyoto Protocol, established at a 1997 meeting, set targets for emission levels in developed countries. A collective 5% reduction in emissions from industrialized countries was <u>mandated</u>.

The problem is very political. Tensions between developed countries (which often achieved their industrialized glory years through processes involving <u>massive</u> emissions of greenhouse gases) and developing countries, which do not want any limits on their potential expansion, are high. In the U.S., there are political tensions over the costs and necessity of expensive emission controls.

Scientifically, the global climate change debate can also be problematic. While there is very strong evidence pointing to the existence of global warming, there is some uncertainty. But how much uncertainty is too much? And how high a level of certainty should be required before action is necessary? In this hour of Science Friday, broadcast live from the annual meeting of the American Association for the Advancement of Science in Washington, D.C., we'll talk about the science, politics, and economics of the global climate change debate. Call in with your comments and questions, and be sure to tune in.

Source: "Global Climate Change," *Talk of the Nation—Science Friday*™, National Public Radio, Inc., February 18, 2000. *www.sciencefriday.com.* Science Friday™ © Samanna Productions, Inc.

 ## What's the Point?

Show your understanding of the reading. Based on the reading, choose the best answer to complete the following statements.

1. The idea that people might cause global climate change was first suggested _____.

   a. in 1992

   b. in the 19th century

   c. after World War II

2. The most commonly known greenhouse gas is _____.

   a. methane

   b. water vapor

   c. carbon dioxide

3. In this summary, global climate change implies that the earth is _____.

   a. getting warmer

   b. getting more rain

   c. losing agricultural areas

4. The issue of global climate change is _____.

   a. purely scientific

   b. highly political

   c. well understood

💡 Understanding Words and Phrases

What is the meaning of the underlined word in each sentence? Answer these questions or do the tasks that follow each sentence. The underlined words are also underlined in the text for easy reference.

1. The archaeologist found <u>fossils</u> of animal bones and plant leaves imprinted in rocks and clay.
   What are some examples of fossil fuels? Why are they called *fossil* fuels?

2. When we use an electric heater, we convert electricity into <u>thermal</u> energy, or heat.
   Can you think of other words that contain the segment *therm?*
   An instrument that measures temperature: _____
   A device that controls temperature in your home: _____
   A container that keeps your drink hot (or cold): _____

3. The government <u>mandated</u> a new energy tax on gasoline.
   What are some things that can be mandated? Who can mandate them?

4. The Great Wall of China is a <u>massive</u> structure that can be seen from far away.
   Give some examples of massive things.

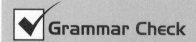Grammar Check

Adjective Phrases

Take a look at the following sentences from the second news summary:

a. The Kyoto Protocol, **established at a 1997 meeting,** set targets for emission levels in developed countries.

b. [Developed countries] often achieved their industrialized glory years through processes **involving massive emissions of greenhouse gases.**

c. While there is very strong evidence **pointing to the existence of global warming,** there is some uncertainty.

What is the function of the words in bold type? What do you notice about their form?

The part of the sentence that is in **bold** type is called an **adjective phrase**. An adjective phrase modifies a noun. An adjective phrase is the result of reducing an *adjective clause*. (In Units 2 and 4, you reviewed adverb clauses and adverb phrases. Recall that a clause is a group of words that has a subject and a verb. A phrase is a group of words that does not have a subject and a verb.) To see this reduction from an adjective clause to an adjective phrase, consider the preceding examples.

a. Sentence with adjective clause:

> The Kyoto Protocol, ***which** **was** **established at a 1997** **meeting,*** set targets for emission levels in developed countries.

b. Sentence with adjective clause:

> [Developed countries] often achieved their industrial-ized glory years through processes ***that** **involve** **massive emissions of greenhouse gases.***

c. Can you write the sentence with the correct adjective clause for example *c* on page 130?

In example *a*, the relative pronoun **which** is also the subject of the adjective clause. Notice that this pronoun is omitted from the adjective phrase. Notice also that the verb **be** is omitted. In example *b*, there is no **be** verb in the clause, and the verb changes to the present participle, or *-ing* form. The resulting adjective phrase modifies the noun *Kyoto Protocol* in *a*, the noun *processes* in *b*, or the noun *evidence* in *c*.

RULES OF REDUCTION:

1. If the adjective clause contains the verb **be**, the relative pronoun and **be** are omitted. This is shown in example *a* on page 130. Here is another example:

   The apples ***that are* in the glass bowl** have been washed.
   → The apples ***in the glass bowl*** have been washed.

2. If the adjective clause contains no **be** verb, the relative pronoun is omitted, and the verb is changed to the *-ing* form. This is shown in examples *b* and *c* on page 130.

3. If the adjective clause is not essential and requires commas, then the resulting phrase also requires commas, as in example *a* on page 130.

4. Only adjective clauses with a *subject* pronoun—**who, which,** or **that**—can be reduced to adjective phrases. For example:

   The two men ***who are* dressed in light-green suits** belong to a special club.
   → The two men ***dressed in light-green suits*** belong to a special club.

   The two men ***(whom) we met earlier*** belong to a special club.
   → cannot be reduced

Exercise 1

In the following sentences, change the adjective clauses to adjective phrases. Which sentences or clauses cannot be changed? Some sentences contain more than one clause.

1. The car which is parked in front of the automobile dealership is a hybrid electric car.

2. Our neighbor's dog barks at everyone who walks by the house.

3. Do you like the restaurant that is on the corner of Salem Street and Prince Street?

4. Oh no! The train that is just leaving the station was ours!

5. Mr. and Mrs. Dean, who are my neighbors, are kind and generous people.

6. The criminal who was guilty of the armed robbery, which was committed two years ago, is now in jail.

7. The Ledyard Bridge, which we are crossing now, was rebuilt several years ago.

8. The Connecticut River, which separates New Hampshire and Vermont, is full of wildlife.

9. All the children who wait in line to see Santa Claus get a lollypop.

10. Yes, this is the museum that I visited two years ago.

11. The patients who stay in this hospital, which is a teaching hospital that was established 200 years ago, are treated by both senior physicians and residents who are in training.

12. Our cat, which we got five years ago, is good at hunting mice that get into the attic.

Exercise 2

In the following sentences, change all adjective phrases to adjective clauses.

1. The woman wearing a straw hat and carrying a briefcase is my chemistry professor.

2. The flowers in our garden, a small and sunny spot in the backyard, are native to this region.

3. The articles published in this magazine were thoroughly reviewed by independent scientists, doing similar research.

4. The athletes, eager to begin the competition, were waiting in the small locker room reserved for visiting teams.

5. Anyone wishing to travel during the busy holiday season should allow extra time needed for security checks.

Exercise 3

For each item, combine the short sentences into one sentence. Use the first sentence as the main clause and the remaining sentences as phrases, if possible, or clauses, if necessary. Use commas if needed.

1. Robertson Davies is one of my favorite authors. He was a well-known Canadian writer.

2. For breakfast, we ate a few pieces of cake. The cake contained dry fruit and nuts.

3. Paul took his girlfriend to a very expensive restaurant. He wanted to impress her. The restaurant was in a beautiful part of town.

4. I still haven't finished reading the book. I started reading it two months ago.

5. This summer, I'm going to drive my new car to visit my cousins. I haven't bought the car yet. My cousins are on the west coast.

 **Let's Talk about It**

1. How are fossil fuels related to carbon dioxide?

2. What is a greenhouse? How is it associated with global warming?

3. Besides the greenhouse effect, what else might cause global warming?

4. Should people do anything about global climate change? What are the reasons for action or inaction?

What Do You Think?

Many people believe that we should reduce industrial emissions and the use of energy to prevent global warming. Are there other reasons for reducing industrial emissions and energy use? Write a short essay explaining your opinions.

Take a look at story summary 3. Does it sound familiar? If not, see Unit 7 in this book!

Story summary 3.

SCIENCE FRIDAY > ARCHIVES > 2000 > FEBRUARY > FEBRUARY 25, 2000

Engineering Feats

from *Talk of the Nation—Science Friday*™,
National Public Radio, Inc.

Think back to the year 1900. What was life like then? Now think of today. See some changes? Ok . . . so how did we get to this point?

This week the National Academy of Engineering listed its top twenty engineering achievements of the century, ranging from electrification to developments in materials science. The basic guideline for inclusion in the list was whether or not an advancement had made a great contribution to the quality of life. (Even devices such as the airplane, invented in the 1800s, were eligible for the list if the judges felt that they had had a great impact on society in the 20th century.) On this hour of Science Friday, we'll talk about the list—and take your suggestions for what YOU think should have been included.

Source: "Engineering Feats," *Talk of the Nation—Science Friday*™, National Public Radio, Inc., February 25, 2000. *www.sciencefriday.com.* Science Friday™ © Samanna productions, Inc.

Fun Factoids

from *Science Friday*™, National Public Radio, Inc.

The following material originally appeared on NPR's website. These are short descriptions of interesting scientific facts, called Fun Factoids. Below are five factoids for you to read and enjoy.

1. Potato Paint

If you're **partial** to ketchup red or hash brown, you might want to check out a new kind of paint . . . that's part potato.

New Scientist magazine reports that chemists in Great Britain have come up with a way to make the brown **tubers** into green paint. They're using plant **starches** to replace up to a quarter of the structural chemicals in standard house paint. In most paints, **vinyl** or **acrylic** plays the role of hardener. But potato starch can do the same job, and it won't even **rot** on the walls.

The chemists say starch will turn out to be <u>considerably</u> cheaper to use than its chemical <u>counterparts</u> . . . and they think potato paint will be a big hit with environmentally aware consumers.

And we say . . . It'll be just the thing . . . especially if you're painting your kitchen.

2. Spinach Explosion

Ah, spinach . . . bane of children . . . beloved of **Popeye** . . . delicious with feta cheese . . . and now . . . scientists say this green leafy vegetable could be used to <u>neutralize</u> powerful explosives.

The Department of Energy's Pacific Northwest Laboratory has been testing various natural **enzymes** and their effect on explosives like TNT. And their researchers say that spinach is just the thing for <u>deflating</u> dynamite, and making blasting caps behave.

It turns out that the spinach enzymes eat and digest the explosives, and <u>transform</u> them into harmless products like carbon dioxide and water. This is actually pretty similar to what takes place in your stomach after dinner.

Scientists <u>extract</u> the enzymes, also found in **fungi** and buttermilk, mix them with a buffer solution, and **lactic acid** or

---

partial: likes especially

tubers: swollen, underground plant stems, such as potatoes, from which new plant shoots grow

starches: natural nutrient carbohydrates, found in plants, especially in potatoes, corn, rice, and wheat

vinyl: a chemical compound used in making plastics

acrylic: a chemical compound used in making some plastics and paint

rot: decay; decompose

Popeye: a famous American cartoon character who likes to eat spinach

enzymes: proteins produced by living organisms and acting as chemical agents

fungi: plants that include yeasts, molds, and mushrooms

lactic acid: a chemical compound present in some foods, like sour milk, and used in chemical processes

ethanol, which <u>concentrates</u> the mix. Add the broth to TNT, wait a little while . . . and voila! the boom is a bust.

You might call it a <u>dynamite</u> recipe for spinach!

ethanol: a type of alcohol that is a type of fuel

3. Airplane Band-Aid

Here's an interesting science note this week . . . the Federal Aviation Administration has given the **thumbs up** to a new band-aid . . . for airplanes.

You may not have realized that jumbo jets get **boo-boos** just like the rest of us. Take-offs, landings, rapid temperature changes and <u>turbulence</u> all take their toll on an airplane's aluminum skin. These daily stresses cause tiny cracks and internal flaws to form, then grow into bigger cracks and bigger problems.

thumbs up (idiom): the okay or approval

boo-boos (informal): small injuries

Airplane doctors used to repair these aircraft by bolting hard metal plates on top of the cracks. But the <u>stiff</u> plates and the bolts were causing new cracks to form.

So the FAA and Sandia National Laboratories came up with a softer, gentler solution: bandages.

The new flexible airplane band-aids are really patches of thin tape . . . and they give aging airplanes the gift of new life . . . virtually overnight. So everybody's happy . . . band-aids save the airlines tens of thousands of dollars.

And you get to concentrate on those gremlins on the wing!

4. Leap Second

How many times have you yelled, "Just a second, just a second!" Well, on June 30th, you will get that second . . . A leap second . . . at exactly 7:59 and 59 seconds eastern time. So make the most of it.

This is the 21st leap second added since 1972. . . . Our guardians of time, the U.S. Naval Observatory, insert these extra seconds to bring universal time into sync with the rotation of the earth. . . . The earth is just not a reliable clock . . . at least when compared to the Observatory's atomic timepieces.

The earth is sort of rotationally challenged . . . its rotation slows down just a little bit . . . about 1 to 3 milliseconds per day per century . . . so the leap second is added to compensate.

It's much easier to change the time on an atomic clock than it is to alter the rotation of the earth . . . unless you're Superman.

5. Mussel Power

Have you heard the news that eating fish is heart-healthy? Well, fish aren't the only sea creatures that may be good for your health.

Scientists at Auckland University in New Zealand are hoping to introduce mussels . . . the kind who live in shells in the ocean that is . . . into the operating room. You know how tough it is to pry one of these things off a rock? Now researchers are synthesizing a **gooey** protein found in the mussels that makes them cling tight. The protein is **secreted** by a gland in the mussel's foot, and it's made up of some mysterious materials. The Auckland chemistry team has spent an entire year trying to <u>assemble</u> a copy of the long chain of amino acids that make up the protein . . . but it should be well worth the trouble.

Doctors could use this super glue to close wounds without **stitches.** If the human body accepts this foreign <u>substance</u>, the adhesive would just eventually <u>dissolve</u> . . . and stitches could become <u>obsolete</u>.

Now that's how to **flex** your mussels!

Source: "Fun Factoids," *Science Friday*™, National Public Radio, Inc., 1998–1999. *www.sciencefriday.com. Science Friday*™ © Samanna Productions, Inc.

gooey: sticky

secreted: produced and separated out from cells or organs

stitches: medically sewing up a wound

flex: contract (a muscle); use muscles in a show of strength (here, a play on words)

What's the Point?

Check your understanding of the texts. Based on the readings, provide short answers to these questions.

1. Potato Paint

   a. How are potatoes used to replace traditional paint?

   b. What are the advantages of potato paint?

   c. Can potato paint be used like regular house paint?

2. Spinach Explosion

   a. What substance from spinach can be used to deactivate explosives?

   b. Where else can we get the same natural substance?

   c. What organization is working to develop this new technology?

Now, make up your own questions about the factoids for your fellow students.

   a. For each of the three remaining factoids, develop two or more questions to test a reader's understanding of the text. They can be short-answer questions, true-or-false statements, or questions with multiple-choice answers. Your teacher will help you come up with appropriate, clear questions.

   b. Exchange your questions with a classmate and answer each other's questions. You can volunteer to pose your questions to the entire class.

 ## Understanding Words and Phrases

Create word puzzles for a classmate. The words and expressions in this exercise appear in the factoids you just read. They are underlined in the texts for your easy reference. Pick six or more words from this list, and write an original sentence using each word. Then, write the sentences on a piece of paper, leaving out the vocabulary words that were on the list. Scramble (mix up) the letters of each missing word and write down the scrambled word next to its sentence. Exchange papers with a classmate and solve the puzzles by unscrambling the words that fit the sentences.

assemble	dissolve	stiff
considerable	dynamite	substance
concentrate	extract	transform
counterpart	neutralize	turbulence
deflate	obsolete	

Example: Honey is a sweet, sticky _____. [abuscents]
Answer: *substance*

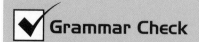

Grammar Check

Infinitives

In Unit 6, we reviewed the use of *gerunds* in certain grammatical situations. Similarly, an **infinitive** (the form *to* + the simple form of a verb) is used in certain other grammatical situations. In this unit, we will review four such situations, or types of uses, which are demonstrated in the Fun Factoids you just read. (There are some additional uses of infinitives that will not be reviewed here.)

Look at these examples from the text. Can you figure out what types of uses require an infinitive? The underlined words or words in parentheses provide a hint.

TYPE 1

You might <u>want</u> ***to check*** out a new kind of paint.

Starch will <u>turn out</u> ***to be*** considerably cheaper to use.

These daily stresses <u>cause</u> tiny cracks and internal flaws ***to form.***

TYPE 2

They're using plant starches (<u>in order</u>) ***to replace . . .*** chemicals in . . . paint.

This . . . vegetable could be used (<u>in order</u>) ***to neutralize*** powerful explosives.

TYPE 3

<u>It's</u> much <u>easier</u> ***to change*** the time on an atomic clock than . . . ***to alter. . . .***

<u>Starch</u> will turn out to <u>be</u> considerably <u>cheaper</u> ***to use.***

(*variation:* <u>It</u> will turn out to <u>be</u> considerably <u>cheaper</u> ***to use*** starch.)

TYPE 4

Chemists . . . have come up with <u>a way</u> **to make** the brown tubers into green paint.

You <u>get</u> (<u>an opportunity</u>) **to concentrate** on those gremlins on the wing!

These common uses of infinitives are summarized:

1. **Type 1, after certain verbs.** Just like gerunds follow certain verbs, infinitives follow certain other verbs. Some verbs may be followed by either a gerund or an infinitive. These verbs must be learned and memorized like vocabulary. With some verbs, there is a noun or pronoun between the verb and infinitive. For example:

   Is Sara <u>planning</u> **to go** home early this afternoon?

   Simon's lawyer <u>advised</u> him **to read** the document carefully.

2. **Type 2, expression of purpose—*in order to*.** This use of infinitives answers the question "why?" The actual words *in order to* are often omitted. For example:

   He goes to the night club (<u>in order</u>) **to meet** people.

3. **Type 3, with *it* as the subject of a sentence.** These sentences usually have the form ***it* + *be/verb/modal* + *adjective* + infinitive.** The subject *it* refers to the infinitive phrase. The adjective describes the infinitive phrase, and the infinitive provides meaningful content to the sentence. (Also, see a variation of this form in the second example in Type 3; the object of the infinitive may be moved to the subject position in the sentence.) For example:

   <u>It is difficult</u> **to design** a computer.
   → *variation*: <u>A computer is difficult</u> **to design.**

   <u>It is easy</u> *for you* **to say,** but not <u>easy</u> *for him* **to do!**

4. **Type 4, after a noun or pronoun that has a particular purpose.**
Here, the infinitive refers to the noun (or pronoun) that precedes it and explains its purpose. Frequent nouns include *a way, a method, a chance, how* (a relative pronoun), *get (an opportunity)* **to do** something. For example:

> Please give me a <u>chance</u> **to prove** my skills.
>
> It takes <u>time</u> **to fix** this problem.
> → *variation*: This problem takes <u>time</u> **to fix.**
>
> She showed me <u>how</u> **to wrap** a gift.

Exercise 1

Here are additional examples of the uses of infinitives from the text of Fun Factoids. Which type of use does each example represent?

1. But the stiff plates and the bolts were causing new cracks **to form.**

2. Our guardians of time, the U.S. Naval Observatory, insert these extra seconds **to bring** universal time into sync with the rotation of the earth. . . .

3. Scientists . . . are hoping **to introduce** mussels . . . into the operating room.

4. [Do] You know how tough it is **to pry** one of these things off a rock? *(Hint: change this question to a statement of fact.)*

5. The leap second is added **to compensate.**

6. The Auckland chemistry team has spent an entire year trying **to assemble** a copy of the long chain of amino acids. . . .

7. Doctors could use this super glue **to close** wounds without stitches.

8. Now that's how **to flex** your mussels!

Exercise 2

In these sentences, choose either an infinitive or gerund form of each verb in parentheses.

1. Ralph can't stand *(hear)* _____ people gossip about their friends or co-workers.

2. You have to see it *(believe)* _____ it!

3. It's fun *(do)* _____ crossword puzzles on a lazy Sunday morning.

4. They encouraged him *(exercise)* _____ every day *(gain)* _____ strength.

5. We look forward to *(see)* _____ you tomorrow *(discuss)* _____ *(expand)* _____ the research project.

6. It can be very soothing *(take)* _____ a hot bath at the end of a stressful day.

7. Jeff has a good reason for *(decide)* _____ *(quit)* _____ his job.

8. There's a new technique *(teach)* _____ children how *(play)* _____ musical instruments.

9. Cars are easier *(drive)* _____ than trucks.

10. Lillian Dooley truly deserves *(win)* _____ the Citizen of the Year Award after *(work)* _____ as hard as she did *(help)* _____ the people in her community.

11. Is it necessary *(look up)* _____ every unknown
word in the dictionary *(understand)* _____ an
article?

12. I usually take the bus to work *(avoid)* _____
(drive) _____ in traffic.

13. Lawrence hates *(get caught)* _____ in the rain, so
he always remembers *(take)* _____ an umbrella
with him.

14. Are dairy farms interesting for children *(visit)* _____?

15. Everyone must begin *(sing)* _____ when I raise
my hand. Continue *(sing)* _____ until I give the
command *(stop)* _____ .

16. If you take Mr. Coleman's class, you'll get *(watch)*
_____ two foreign films.

Let's Talk about It

1. When you read the "fun factoids," did any question arise in your
mind? Do you have any comments about what you just read? Did
anything remind you of a related issue?

   Think of a question or issue that you would like to discuss in class. It
   can be related to any of the "factoids" you read. Pose your question
   or issue to the whole class for discussion.

2. If possible, listen to a radio show or watch a TV show in class. Then,
discuss the show. What did you learn? (NPR tapes and transcripts
are available for purchase. Tapes and transcripts are also available
for some of the public TV shows. For details, call your local radio
and TV stations or check their websites.)

What Do You Think?

Choose a question or issue from the activity on page 144—presented either by you or by a classmate—and write a brief essay describing your response or your thoughts.

Expansion Activities

1. Play a game of knowledge! Websites sometimes have interactive programs. One such activity that appeared on the *Science Friday*™ website was the Science Quiz. A few of the quiz questions are included for you on page 146. See if you know the answers, which are provided on the page following the quiz.

   After taking the quiz, discuss these questions with your classmates:

   a. Do you know of any other foods, drinks, or herbs that can improve your health?
   b. Ants help other ants find food by leaving a trail. Do people ever do anything similar?
   c. Have you ever had a bad reaction to contact with certain plants or animals?

2. Visit the NPR, *Science Friday*™, or another program website. Find something that interests you and read about it. You can read summaries of news reports and science shows or do other activities. You can listen to a complete show right on the computer (for this, you may have to download a free audio software program). Report the news to your class, or bring an interesting activity to class.

3. Listen to one of the radio shows suggested in this unit. Or watch one of the TV shows. Tell your class what you have learned.

Science Quiz

from *Science Friday*™, National Public Radio, Inc.

It's time for another SOUNDS LIKE SCIENCE quiz!

Question Number One

In honor of the Fourth of July holiday, we are going on a picnic and of course we brought some hamburgers to grill. They're delicious but not what many would call a health food. Recently scientists say they found a way to make hamburgers better for you—by adding which of these ingredients?

    a. Sesame oil
    b. Orange Juice
    c. Cherries
    d. A touch of cilantro

Great! Let's move on to number two.

Question Two

Like any respectable picnic, we've attracted a few ants. One of the many interesting things that ants can do is conveniently find the shortest path to the food. How do they do it—with which sense?

    a. sight
    b. touch
    c. smell

And how 'bout another one.

Question Three

Another picnic danger is poison ivy. If you have found yourself highly susceptible to the plant before, you should also avoid:

    a. Ginkgo nuts
    b. Lacquered boxes
    c. Cashew shells
    d. All of the above

Answers

1c.

The answer is Cherries. Scientists at Michigan State University say that adding cherries to hamburger retards spoilage and reduces the risk of cancer. The cherry tissue slows down the oxidative deterioration of meat lipids, and reduces the formation of suspected cancer-causing compounds known as H-A-As or Heterocyclic Aromatic Amines. Cherry tissue has also been shown to make a hamburger that's lower in fat and more tender than your average beef variety.

2c.

Smell is the answer. It's not their great mind; ants only have a few hundred neurons to help them work out what to do next. What they do have are pheromones. When an ant comes across food, it lays down a pheromone trail for other ants to follow. At first other ants will follow long and short paths at random. Eventually though more ants will travel the shorter path in a given time and will create the stronger pheromone trail. This will become the favored path.

3d.

The answer is All of the above. The substance in poison ivy that causes a rash can also be found in other plants such as the cashew tree, the ginkgo tree, and the Japanese lacquer tree. Some people have developed rashes after coming in contact with lacquered products form Japan or China, or cashew nut shells or gingko nut shells.

Source: "Science Quiz," *Science Friday*™, National Public Radio, Inc. 2000. *www.sciencefriday.com.* Science Friday™ © Samanna Productions, Inc.

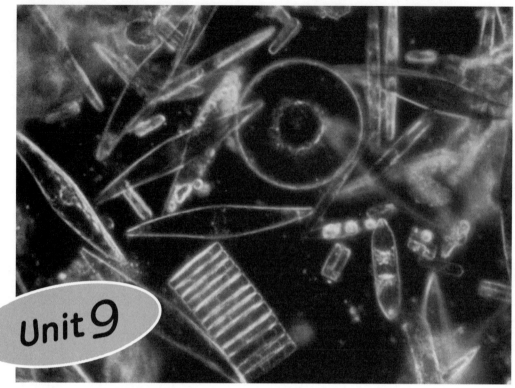

Marine diatoms as seen through a microscope.

Our Future

Search Your Knowledge

1. Imagine life 50 years ago. What products, services, and conveniences do we have today that we didn't have then?

2. What are scientists and engineers working on now that might change our future? Think of articles you have read, news you have heard, and your own experiences.

3. Make some predictions about the future. What new products and services might we enjoy 50 years from now? List some possible future innovations in the following fields:

> Medicine

What is done for diseased joints today? What could be done in the future?

> Energy

How is solar energy used today? How can its use be improved?

> Computing

Name some major innovations in computer technology in the last 20 years. What are some current areas of research or development?

> Software

What are some annoying problems with many computer programs? How can we make software more dependable?

> Communications

Why is secrecy important in communications? How can we achieve it, particularly in Internet-based communications?

The text that you are about to read appeared in the February 2003 issue of *Technology Review* magazine. It predicts technological advances that will greatly influence our future. Do you agree with these predictions?

The original article, titled "10 Emerging Technologies That Will Change the World," was divided into an introduction and ten stories, each story describing a different technology. This unit contains the article introduction and five of the technology stories, each in complete form and in a separate section. Here are the contents of this unit:

Introduction
Section 1: "Injectable Tissue Engineering" by Alexandra M. Goho
Section 2: "Nano Solar Cells" by Eric Scigliano
Section 3: "Grid Computing" by M. Mitchell Waldrop
Section 4: "Software Assurance" by Erika Jonietz
Section 5: "Quantum Cryptography" by Herb Brody

The full article excerpt is long, but it is conveniently divided into shorter sections. Read one section at a time and complete the corresponding reading-comprehension and vocabulary exercises after each section. Can you understand the underlined words in the text? These words may be new to you; therefore, they are underlined so that you can find them easily later on, if you wish to refer to them again. See if you can figure out what they mean from context or from the other words and meanings around the underlined word. The words are also included in the vocabulary exercises under Key Words and Understanding Words and Phrases.

The boldfaced words in the text are glossed in the margin. These non–high frequency vocabulary words or phrases are helpful to understanding the reading.

KEY WORDS

The key words in this unit are divided into separate sections: the article introduction and the five story sections. Familiarize yourself with these words before reading each section. The key words are underlined in the text you are about to read.

INTRODUCTION

Use these sentences to figure out the meanings of the underlined words.

1. After sleeping there all winter, the bear finally underlined emerged from the cave in the spring.
   Give examples of some things that emerge from the ground.

2. Rice is an important crop in East Asia.
   Name some major crops grown in the United States.

3. A can opener is a very useful gadget.
   What gadgets do you have in your kitchen?

SECTION 1: Injectable Tissue Engineering

Figure out the meanings of the underlined words.

1. Some drugs are taken by mouth, while others are injected through a needle under the skin.
   Name some things that you can inject into other things.

2. Skin, bones, and leaves on trees are all made up of different living tissues.

   Give some other examples of living tissue.

3. All living tissues are made up of tiny, individual living cells.

   What are the main structural parts of a cell? What organisms consist of a single cell?

4. A syringe is used to inject, squeeze out, or even suck in a substance. It consists of a cylindrical container and a hollow needle or nozzle.

   What are some different uses for syringes?

5. Cartilage is a strong, flexible, connective tissue found in joints, such as knees and hips.

   Your nose and ears are made of cartilage. Where else in your body can you find it?

6. A cadaver is a dead human body, especially one that is used for medical training.

   Most cadavers used in medical schools come from people who donate their bodies.

SECTION 2: Nano Solar Cells

Analyze the meanings of these words and answer the related questions.

1. *photovoltaic:* (adjective) capable of producing a voltage when exposed to light

   What are photovoltaic cells?

2. *composite:* (noun) a complex material made of two or more distinct substances, which combine to produce special properties, such as structural strength

   Concrete, a strong material used in construction, is a kind of composite; it consists of sand, broken stone, pebbles, and cement. Can you think of other examples of composite materials, or things made from composites?

3. *electrode:* (noun) a solid conductor, usually metal, through which electric current enters or leaves an electrically conducting medium, such as a battery or solar cell

Locate the electrodes on a small battery, which may be used in a radio, a smoke detector, or a toy. The electrodes are often labeled (+) and (−).

SECTION 3: Grid Computing

Figure out the meanings of the underlined words. Then answer the related questions.

1. A <u>protocol</u> is a code of correct conduct; it can be a detailed plan for a medical procedure or scientific experiment. In computer science, a <u>protocol</u> is a standard procedure for regulating the flow of data between computers.

   Why is it important to have protocols in certain professions or for certain actions?

2. On Labor Day, a large crowd gathered around a smoky barbecue grill in the town park. By evening, the last hamburger was cooked, and the smoke from the grill <u>dispersed</u>. This was followed by <u>dispersal</u> of the crowd. By sunset, the park was empty.

   Give two other examples of things that can disperse.

3. While we were watching a movie in a theater, the film projector began to burn! The manager, who was worried about a fire, immediately told everyone to <u>evacuate</u> the building.

   Have you ever been in an emergency situation where <u>evacuation</u> was required?

4. Only a small number of experts truly understand the <u>arcane</u> U.S. tax laws.

   What do you think *arcane* means?

5. McDonald's fast-food restaurants are <u>ubiquitous</u> not only throughout the United States but throughout the world.

   Name something else that is ubiquitous (either locally or in a larger area).

SECTION 4: Software Assurance

Here are some expressions that will help you understand this section.

1. *to crash:* — in computer science: to stop working suddenly, often due to a sudden failure of a computer program

2. *software bug:* — a mistake or problem in a computer program

3. *to reboot:* — to turn off (a computer) and turn it on again; to restart

4. *civil engineer:* — an engineer who works on designing and constructing public works, such as bridges, roads, dams, sewage systems, other large facilities

5. *pseudo:* — a prefix (or adjective) meaning false, deceptive, similar but not real. For example, the author Samuel Clemens used the *pseudo*nym Mark Twain.

SECTION 5: Quantum Cryptography

Do you know what these words mean? They are useful for understanding this story.

1. *cryptography:* — the process of communicating in secret writing (or code); from *crypt-*, meaning hidden or secret

   encryption: — the act of putting into secret code; encoding

   decryption: — the act of solving a secret code; decoding

2. an *entrepreneur:* — a person who starts or operates a business, especially a new and financially risky business

   Can you name some well-known entrepreneurs who became successful?

3. *to crack* (a code) or *to break* (a code): — to figure out a solution to a difficult problem, especially after a lot of effort

   Why do people use secret codes for communication? Why do others want to crack (or break) the codes?

10 <u>Emerging</u> Technologies That Will Change the World

from *Technology Review*

In labs around the world, researchers are busy creating technologies that will change the way we conduct business and live our lives. These are not the latest <u>crop</u> of <u>gadgets</u> and **gizmos:** they are completely new technologies that could soon transform computing, medicine, manufacturing, transportation, and our energy infrastructure. . . . In this special issue, *Technology Review*'s editors have identified 10 <u>emerging</u> technologies that we predict will have a tremendous influence in the near future. For each, we've chosen a researcher or research team whose work and vision is driving the field. The profiles . . . offer a **sneak preview** of the technology world in the years and decades to come.

> **gizmos:** mechanical devices or parts whose names are unknown

> **sneak preview:** an advance look

Section 1: <u>Injectable Tissue</u> Engineering

Jennifer Elisseeff

Every year, more than 700,000 patients in the United States <u>undergo</u> joint replacement surgery. The procedure—in which a knee or a hip is replaced with an artificial <u>implant</u>—is highly <u>invasive</u>, and many patients delay the surgery for as long as they can. Jennifer Elisseeff, a biomedical engineer at Johns Hopkins University, hopes to change that with a <u>treatment</u> that **does away with** surgery entirely: injectable <u>tissue</u> engineering. She and her colleagues have developed a way to <u>inject</u> joints with specially designed mixtures of polymers, <u>cells</u>, and growth stimulators that solidify and form healthy <u>tissue</u>. "We're not just trying to improve the current therapy," says Elisseeff. "We're really trying to change it completely."

> **does away with** (phrasal verb): eliminates or removes

Elisseeff is part of a growing movement that is pushing the <u>bounds</u> of <u>tissue</u> engineering—a field researchers have long hoped would produce lab-grown alternatives to <u>transplanted</u> organs and <u>tissues</u>. For the last three decades, researchers have focused on growing new tissues on polymer scaffolds in the lab. While this approach has had success producing small amounts of cartilage and skin, researchers have had difficulty keeping cells alive on larger scaffolds. And even if those problems could be worked out, surgeons would still have to implant the lab-grown tissues. Now, Elisseeff, as well as other academic and industry researchers, are turning to injectable systems that are less invasive and far cheaper. Many of the tissue-engineering applications to reach the market first could be delivered by syringe rather than implants, and Elisseeff is pushing to make this happen as soon as possible.

Elisseeff and her colleagues have used an injectable system to grow cartilage in mice. The researchers added cartilage cells to a light-sensitive liquid polymer and injected it under the skin on the backs of mice. They then shone ultraviolet light through the skin, causing the polymer to harden and encapsulate the cells. Over time, the cells multiplied and developed into cartilage. To test the feasibility of the technique for minimally invasive surgery, the researchers injected the liquid into the knee joints of cadavers. The surgeons used a fiber-optic tube to view the hardening process on a television monitor. "This has huge implications," says James Wenz, an orthopedic surgeon at Johns Hopkins who is collaborating with Elisseeff.

While most research on injectable systems has focused on cartilage and bone, observers say this technology could be extended to tissues such as those of the liver and heart. The method could be used to replace diseased portions of an organ or to enhance its functioning, says Harvard University pediatric surgeon Anthony Atala. In the case of heart failure, instead of opening the chest and surgically implanting an engineered valve or muscle tissue, he says, simply injecting the right combination of cells and signals might do the trick.

For Elisseeff and the rest of the field, the next frontier lies in a powerful new tool: stem cells. Derived from sources like bone marrow and embryos, stem cells have the ability to differentiate into numerous types of cells. Elisseeff and her colleagues have exploited that ability to grow new cartilage and bone simultaneously—one of the trickiest feats in tissue engineering. They made layers of a polymer-and-stem-cell mixture, infusing each layer with specific chemical signals that triggered the cells to develop into either bone or cartilage. Such hybrid materials would simplify knee replacement surgeries, for instance, that require surgeons to replace the top of the shin bone and the cartilage above it.

Others in Injectable Tissue Engineering	
RESEARCHER	**PROJECT**
Anthony Atala Harvard Medical School	Cartilage
Jim Burns Genzyme	Cartilage
Antonios Mikos Rice University	Bone and cardiovascular <u>tissue</u>
David Mooney University of Michigan	Bone and cartilage

Don't expect <u>tissue</u> engineers to grow entire artificial organs anytime soon. Elisseeff, for one, is aiming for smaller advances that will make <u>tissue</u> engineering a reality within the decade. For the thousands of U.S. patients who need new joints every year, such small feats could be huge.

—*Alexandra M. Goho*

Section 2: Nano Solar Cells
Paul Alivisatos

The sun may be the only energy source big enough to **wean us off** fossil fuels. But <u>harnessing</u> its energy depends on silicon wafers that must be produced by the same <u>exacting</u> process used to make computer chips. The expense of the silicon wafers raises <u>solar-power</u> costs to as much as 10 times the price of fossil fuel generation—keeping it an energy source best <u>suited</u> for satellites and other niche applications.

wean us off: make us give up a habit (verb: *to wean off*)

Paul Alivisatos, a chemist at the University of California, Berkeley, has a better idea: he aims to use **nano**technology to produce a <u>photovoltaic</u> material that can be spread like plastic wrap or paint. Not only could the nano solar cell be integrated with other building materials, it also offers the promise of cheap production costs that could finally make solar power a widely used electricity alternative.

nano: one billionth of a measure

Alivisatos's approach begins with electrically conductive polymers. Other researchers have attempted <u>to concoct</u> solar cells from these plastic materials . . . but even the best of these devices aren't nearly efficient enough at converting solar energy into electricity. To improve the efficiency, Alivisatos and his coworkers are adding a new ingredient to the polymer: nanorods, bar-shaped semiconducting inorganic crystals measuring just seven nanometers by 60 nanometers. The result is a cheap and flexible material that could provide the same kind of efficiency achieved with silicon solar cells. Indeed, Alivisatos hopes that within three years, Nanosys—a Palo Alto, CA, startup he cofounded—will roll out a nanorod solar cell that can produce energy with the efficiency of silicon-based systems.

The prototype solar cells he has made so far consist of sheets of a nanorod-polymer <u>composite</u> just 200 nanometers thick. Thin layers of an <u>electrode</u> sandwich the composite sheets. When sunlight hits the sheets, they absorb **photons,** exciting electrons in the polymer and the nanorods, which make up 90 percent of the composite. The result is a useful current that is carried away by the <u>electrodes</u>.

photons (physics): a unit of electromagnetic energy considered a separate particle

Early results have been encouraging. But several tricks now **in the works** could further <u>boost</u> performance. First, Alivisatos and his collaborators have switched to a new nanorod material, cadmium telluride, which absorbs more sunlight than cadmium selenide, the material they used initially. The scientists are also <u>aligning</u> the nanorods in branching assemblages that conduct electrons more efficiently than do randomly mixed nanorods. "It's all a matter of processing," Alivisatos explains, adding that he sees "no inherent reason" why the nano solar cells couldn't eventually match the performance of <u>top-end</u>, expensive silicon solar cells.

in the works: under development

The nanorod solar cells could be rolled out, inkjet printed, or even painted onto surfaces, so "a billboard on a bus could be a solar collector," says Nanosys's director of business development, Stephen Empedocles. He predicts that cheaper materials could create a $10 billion annual market for solar cells, <u>dwarfing</u> the growing market for conventional silicon cells.

Alivisatos's nanorods aren't the only technology entrants chasing cheaper solar power. But whether or not his approach eventually revolutionizes solar power, he is bringing novel nanotechnology strategies to bear on the problem. And that alone could be a major contribution to the search for a better solar cell. "There will be other research groups with clever ideas and processes—maybe something we haven't even thought of yet," says Alivisatos. "New ideas and new materials have opened up a period of change. It's a good idea to try many approaches and see what emerges."

Others in Nano Solar Cells	
RESEARCHER	**PROJECT**
Richard Friend University of Cambridge	Fullerene-polymer composite solar cells
Michael Grätzel Swiss Federal Institute of Technology	Nanocrystalline dye-sensitized solar cells
Alan Heeger University of California, Santa Barbara	Fullerene-polymer composite solar cells
N. Serdar Sariciftci Johannes Kepler University	Polymer and fullerene-polymer composite solar cells

Thanks to nanotechnology, those new ideas and new materials could transform the solar cell market from a boutique source to the Wal-Mart of electricity production.

—*Eric Scigliano*

Section 3: Grid Computing
Ian Foster & Carl Kesselman

In the 1980s "internetworking <u>protocols</u>" allowed us to link any two computers, and a vast network of networks called the Internet exploded around the globe. In the 1990s the "hypertext transfer <u>protocol</u>" allowed us to link any two documents, and a vast, online library-**cum**-shopping-mall called the World Wide Web exploded across the Internet. Now, fast-emerging "grid <u>protocols</u>" might allow us to link almost anything else: databases, simulation and visualization tools, even the <u>number-crunching</u> power of the computers themselves. And we might soon find ourselves in the midst of the biggest explosion yet.

cum (Latin): with

"We're moving into a future in which the location of [computational] resources doesn't really matter," says Argonne National Laboratory's Ian Foster. Foster and Carl Kesselman of the University of Southern California's Information Sciences Institute <u>pioneered</u> this concept, which they call grid computing in analogy to the electric grid, and built a community to support it. Foster and Kesselman, along with Argonne's Steven Tuecke, have led development of the Globus Toolkit, an open-source <u>implementation</u> of grid protocols that has become the **de facto** standard. Such protocols

de facto (Latin): in reality; actually

promise to give home and office machines the ability to reach into cyber-space, find resources wherever they may be, and assemble them on the fly into whatever applications are needed.

Imagine, says Kesselman, that you're the head of an emergency response team that's trying to deal with a major chemical spill. "You'll probably want to know things like, What chemicals are involved? What's the weather fore-cast, and how will that affect the pattern of <u>dispersal</u>? What's the current traffic situation, and how will that affect the <u>evacuation</u> routes?" If you tried to find answers on today's Internet, says Kesselman, you'd **get bogged down** in <u>arcane</u> log-in procedures and <u>incompatible</u> software. But with grid computing it would be easy: the grid proto-cols provide standard mechanisms for discovering, accessing, and <u>invoking</u> just about any online resource, simultaneously building in all the requisite safeguards for security and authentication.

get bogged down: be held back

Construction is **under way** on dozens of distributed grid com-puters around the world—<u>virtually</u> all of them employing Globus Toolkit. They'll have <u>unprecedented</u> computing power and applica-tions ranging from genetics to particle physics to earthquake engineering. The $88 million TeraGrid of the U.S. National Science Foundation will be one of the largest. When it's completed later this year, the general-purpose, dis-tributed supercomputer will be capable of <u>some</u> 21 trillion floating-point operations per second, making it one of the fastest computational systems on Earth. And grid computing is experiencing an <u>upsurge</u> of support from industry <u>heavyweights</u> such as IBM, Sun Microsystems, and Microsoft. IBM, which is a primary partner in the TeraGrid and several other grid projects, is beginning to market an enhanced commercial version of the Globus Toolkit.

under way: happening now

Others in Grid Computing	
RESEARCHER	PROJECT
Andrew Chien Entropia	Peer-to-Peer Working Group
Andrew Grimshaw Avaki; University of Virginia	Commercial grid software
Miron Livny University of Wisconsin, Madison	Open-source system to harness idle workstations
Steven Tuecke Argonne National Laboratory	Globus Toolkit

Out of Foster and Kesselman's work on protocols and standards, which began in 1995, "this entire grid movement emerged," says Larry Smarr, director of the California Institute for Telecommunications and Information Technology. What's more, Smarr and others say, Foster and Kesselman have been <u>instrumental</u> in building a community around grid computing and in <u>advocating</u> its integration with two related approaches: peer-to-peer computing, which brings to bear the power of <u>idle</u> desktop computers on big problems in the manner made famous by SETI@home, and Web services, in which access to **far-flung** computational resources is provided through enhancements to the Web's hypertext protocol. By helping to merge these three powerful movements, Foster and Kesselman are bringing the grid revolution much closer to reality. And that could mean <u>seamless</u> and <u>ubiquitous</u> access to unfathomable computer power.

far-flung: wide-ranging; distant

—*M. Mitchell Waldrop*

Section 4: Software Assurance
Nancy Lynch & Stephen Garland

Computers <u>crash</u>. That's a fact of life. And when they do, it's usually because of a <u>software bug</u>. Generally, the <u>consequences</u> are minimal—a muttered curse and a <u>reboot</u>. But when the software is running complex distributed systems such as those that support air traffic control or medical equipment, a <u>bug</u> can be very expensive, and even cost lives. To help avoid such disasters, Nancy Lynch and Stephen Garland are creating tools they hope will <u>yield</u> nearly error-free software.

Working together at MIT's Laboratory for Computer Science, Lynch and Garland have developed a computer language and programming tools for making software development more <u>rigorous</u>, or as Garland puts it, to "make software engineering more like an engineering discipline." <u>Civil engineers</u>, Lynch points out, build and test a model of a bridge before anyone constructs the bridge itself. Programmers, however, often start with a goal and, perhaps after some discussion, simply sit down to write the software code. Lynch and Garland's tools allow programmers to model, test, and <u>reason</u> about software before they write it. It's an approach that's unique among efforts <u>launched</u> recently by the likes of Microsoft, IBM, and Sun Microsystems to improve software quality and even to simplify and improve the programming process itself.

Like many of these other efforts, Lynch and Garland's approach starts with a concept called abstraction. The idea is to begin with a **high-level** summary of the goals of the program and then write a series of progressively more specific statements that describe both steps the program can take to reach its goals and how it should perform those steps. For example, a high-level abstraction for an aircraft <u>collision</u> avoidance system might specify that <u>corrective action</u> take place whenever two planes are flying too close. A **lower-level** design might have the aircraft exchange messages to determine which should <u>ascend</u> and which should <u>descend</u>.

high-level: overall or general level of a system

lower-level: specific or detailed level of a system

Lynch and Garland have taken the idea of abstraction further. A dozen years ago, Lynch developed a mathematical model that made it easier for programmers to tell if a set of abstractions would make a distributed system behave correctly. With this model, she and Garland created a computer language programmers can use to write "<u>pseudo</u>code" that describes what a program should do. With his students, Garland has also built tools to prove that lower levels of abstractions relate correctly to higher levels and to simulate a program's behavior before it is translated into an actual programming language like Java. By directing programmers' attention to many more possible bug-<u>revealing</u> circumstances than might be checked in typical software tests, the tools help assure that the software will always work properly. Once software has been thus tested, a human can easily translate the <u>pseudo</u>code into a standard programming language.

Not all computer scientists agree that it is possible to prove software error free. Still, says Shari Pfleeger, a computer scientist for RAND in Washington, DC, mathematical methods like Lynch and Garland's have a place in software design. "Certainly using it for the most <u>critical</u> parts of a large system would

Others in Software Assurance	
RESEARCHER	**PROJECT**
Gerard Holzmann Bell Labs	Software to detect bugs in networked computers
Charles Howell Mitre	Benchmarks for software assurance
Charles Simonyi Intentional Software	Programming tools to improve software
Douglas Smith Kestrel Institute	Mechanized software development

be important, whether or not you believe you're getting 100 percent of the problems out," Pfleeger says.

While some groups have started working with Lynch and Garland's software, the duo is pursuing a system for automatically generating Java programs from highly specified pseudocode. The aim, says Garland, is to "cut human interaction to near zero" and eliminate transcription errors. Collaborator Alex Shvartsman, a University of Connecticut computer scientist, says, "A tool like this will take us slowly but surely to a place where systems are much more dependable than they are today." And whether we're boarding planes or going to the hospital, we can all appreciate that goal.

—*Erika Jonietz*

Section 5: Quantum Cryptography
Nicolas Gisin

The world runs on secrets. Governments, corporations, and individuals—to say nothing of Internet-based businesses—could scarcely function without secrecy. Nicolas Gisin of the University of Geneva is in the vanguard of a technological movement that could fortify the security of electronic communications. Gisin's tool, called quantum cryptography, can transmit information in such a way that any effort to **eavesdrop** will be detectable.

to eavesdrop: to secretly listen to conversations of other people

The technology relies on quantum physics, which applies at atomic dimensions: any attempt to observe a quantum system inevitably alters it. After a decade of lab experiments, quantum cryptography is approaching feasibility. "We can now think about using it for practical purposes," says Richard Hughes, a quantum cryptography pioneer at the Los Alamos National Laboratory in New Mexico. Gisin—a physicist and entrepreneur—is leading the charge to bring the technology to market.

The company that Gisin spun off from his University of Geneva laboratory in 2001, id Quantique, makes the first commercially available quantum-cryptography system, he says. The PC-size prototype system includes a random-number generator (essential for creating a decryption key) and devices that emit and detect the individual photons of light that make up the quantum signal.

Conventional cryptographers concentrate on developing strong digital locks to keep information from falling into the wrong hands. But even the strongest lock is useless if someone steals the key. With quantum cryptography, "you can be certain that the key is secure," says Nabil Amer, manager of the physics of information group at IBM Research. Key transmission takes the form of photons whose direction of polarization varies randomly. The sender and the intended recipient compare polarizations, photon by photon. Any attempt to tap this signal alters the polarizations in a way that the sender and the intended recipient can detect. They then transmit new keys until one gets through without disturbance.

Quantum cryptography is still ahead of its time. Nonquantum <u>encryption</u> schemes such as the public-key systems now commonly used in business have yet to be <u>cracked</u>. But the security of public-key systems relies on the inability of today's computers to work fast enough to <u>break the code</u>. Ultimately, as computers get faster, this defense will <u>wear thin</u>. Public-key encryption, Gisin says, "may be good enough today, but someone, someday, will find a way to <u>crack</u> it. Only through quantum cryptography is there a guarantee that the coded messages sent today will remain secret forever."

Gisin has no <u>illusions</u> about the challenges he faces. For one thing, quantum cryptography works only over the distance a light pulse can travel through the air or an optical fiber without a boost; the process of amplification destroys the quantum-encoded information. Gisin's team holds the world's distance record, having transmitted a quantum key over a 67-kilometer length of fiber connecting Geneva and Lausanne, Switzerland.

Others in Quantum Cryptography	
RESEARCHER	**PROJECT**
Nabil Amer IBM	Quantum key exchange through optical fiber
Richard Hughes Los Alamos National Laboratory	Ground-to-satellite optical communications
John Preskill **Caltech**	Quantum information theory
John Rarity QinetiQ	Through-air quantum-key transmission
Alexei Trifonov and Hoi-Kwong Lo MagiQ Technologies	Quantum-cryptography hardware

Caltech: California Institute of Technology

The work of Gisin and others could usher in a new epoch of quantum information technology. Ironically, it is in part the <u>prospect</u> that superfast quantum computers will someday supply fantastic code-<u>breaking</u> power that drives Gisin and others to perfect their method of <u>sheltering</u> secret information. In the coming decades, Gisin contends, "e-commerce and e-government will be possible only if quantum communication widely exists." Much of the technological future, in other words, depends on the science of secrecy.

—*Herb Brody*

Source: "10 Emerging Technologies That Will Change the World," *Technology Review,* February 2003. Copyright © 2003 Technology Review.

What's the Point? Section 1

Check your understanding of the text. Based on the reading, decide whether the following statements are true (T) or false (F). Some of these ideas are stated directly in the text, while others are implied.

1. _____ Injectable tissue engineering is intended to replace surgery.

2. _____ With injectable tissue engineering, patients would receive treatment sooner than they would undergo joint replacement surgery.

3. _____ This new treatment would be more expensive than transplant surgery.

4. _____ With this new technology, tissues will be grown in laboratories first, and then transplanted into bodies.

5. _____ Injectable tissue engineering would be less painful and less risky than surgery.

6. _____ Biomedical engineers have not yet succeeded in growing cartilage in a living human body.

7. _____ So far, most researchers in this field have concentrated their efforts on the heart.

8. _____ This technique could be used to repair failing parts of organs.

9. _____ This new technology has the advantage of replacing different kinds of tissue.

10. _____ In the next ten years, scientists expect to grow complete organs using this technology.

Understanding Words and Phrases: Section I

Choose one of the words or phrases to fill in the blank in each of the following sentences. You may have to change the form of the word to fit the sentence.

bounds	frontier	scaffold
enhance	implant	transplant
exploit	infuse	treatment
feasibility	invasive	undergo

1. Last year, Margaret _____ a series of tests to evaluate her fitness for space travel.

2. Because Alex was poor when he was growing up, the habits of living modestly and never wasting food are firmly _____ in his character.

3. They had four trees in their front yard and no trees in the back-yard, so their gardener suggested that they _____ two of the trees to the back.

4. When they painted the outside of the five-story building, the painters supported themselves on a large _____, which was set up temporarily along the walls.

5. "Don't worry," said the worker. "Fixing the roof will be much less _____ than painting the classrooms. We won't interrupt your class; we won't even enter the school building."

6. A common _____ of cancer is a combination of surgery, radiation, and drugs.

7. The soil in our back yard is very rich, so we _____ it by growing a wide variety of vegetables in our garden.

8. We are not allowed to camp outside the _____ of the campground.

9. We have explored much of our land as well as our solar system, but the deep ocean is still largely an unexplored _____.

10. Rebuilding the town after damage by the tornado will not be _____ without help from the government.

11. At the end of the delicious Indian dinner, we enjoyed sweet tea _____ with aromatic spices.

12. Unfortunately, the top two Olympic swimmers were disqualified from the competition because they used drugs _____ their performance.

 ## What's the Point? Section 2

Show your understanding of the reading. Based on the text, choose the best answer to complete the following statements.

1. The reason that solar power is not used widely today is that _____.

   a. current solar technology cannot produce enough electricity
   b. solar energy is more expensive to use than other forms of energy
   c. solar power is not available everywhere in the world
   d. solar energy is unpopular

2. Wider use of solar power may now be possible because _____.

    a. new materials have been developed

    b. new production techniques have been developed

    c. a new kind of solar cell can be integrated with other building materials

    d. All of the above are correct.

3. Solar cells convert sunlight into useful energy in the form of _____.

    a. intense heat

    b. bright light

    c. invisible radiation

    d. electric current

4. The new nano solar cells will be made of _____.

    a. a plastic-based material

    b. a silicon-based material

    c. a very thin metal material

    d. All of the above are correct.

5. The composite material of nano solar cells is _____ semi-conducting nanorods.

    a. half

    b. less than half

    c. more than half

    d. None of the above is correct.

6. Alivisatos and his research team are now working to improve the efficiency of nano solar cells with _____.

    a. a new nanorod material and orderly placement of nanorods in the composite

    b. a new nanorod material and random mixing of nanorods in the composite

    c. random mixing of nanorods and painting the solar cells on a surface

    d. orderly placement of nanorods and thinly rolled sheets of solar cells

7. With nanotechnology, solar cells can be _____.

   a. efficient but difficult to produce

   b. expensive but still profitable

   c. inexpensive and widely available

   d. inexpensive but unpopular

 Understanding Words and Phrases: Section 2

Following are excerpts from the text you just read. Based on context, choose the correct meaning of the underlined expression. The underlined expressions appear here in the same order as in the text; they are also underlined in the text for easy reference.

1. "But <u>harnessing</u> its energy depends on silicon wafers. . . ."

   a. controlling and using

   b. buying and selling

   c. creating and destroying

   d. raising and lowering

2. ". . . silicon wafers that must be produced by the same <u>exacting</u> process used to make computer chips."

   a. exactly the same

   b. scientifically advanced

   c. needing pure water

   d. requiring great care

3. ". . . an energy source best <u>suited</u> for satellites and other niche applications."

   a. invented

   b. rejected

   c. inexpensive

   d. appropriate

4. "Other researchers have attempted <u>to concoct</u> solar cells from these plastic materials. . . ."

    a. to prepare by mixing

    b. to invent

    c. to shape into a cone

    d. to cut out

5. "But several tricks now in the works could further <u>boost</u> performance."

    a. inform

    b. increase

    c. bring back

    d. damage

6. "The scientists are also <u>aligning</u> the nanorods in branching assemblages that conduct electrons more efficiently than do randomly mixed nanorods."

    a. making bigger

    b. marketing abroad

    c. manufacturing

    d. arranging in proper order

7. ". . . he sees 'no inherent reason' why the nano solar cells couldn't eventually match the performance of <u>top-end</u>, expensive silicon solar cells."

    a. upside down

    b. among the best

    c. difficult to find

    d. most complex

8. "He predicts that cheaper materials could create a $10 billion annual market for solar cells, <u>dwarfing</u> the growing market for conventional silicon cells."

    a. making seem small

    b. collaborating with

    c. destroying magically

    d. helping cheerfully

What's the Point? Section 3

Check your understanding of the text. Based on the reading, complete the following sentences with brief statements. Your statements should be based on the article, but use your own words.

1. The name *Internet* comes from _____
   _____.

2. Creation of the Internet became possible because people were able to link together _____
   _____.

3. The World Wide Web became possible when _____
   _____.

4. Grid computing would be possible if we could connect _____
   _____.

5. The term *grid computing* comes from _____
   _____.

6. The concept of grid computing was introduced by two scientists who work at _____
   _____.

7. New software protocols, such as the Globus Toolkit, will allow personal computers to _____
   _____.

8. Today's Internet cannot accomplish the things that Globus Toolkit can because _____
   _____.

9. Possible applications for grid computing include _____
   _____.

10. Grid computing has a good chance to succeed commercially due to interest from _____
    _____.

11. Foster and Kesselman are collaborating with others and hope to combine grid computing with two other methods, called
    _____ and _____.

💡 Understanding Words and Phrases: Section 3

I. The story you just read contains some idioms and metaphors that are listed on the right. Some of them are underlined in the text for your reference; others are defined in the margin. These expressions are also answers to riddles that appear on the left. Solve these riddles by matching each item on the left with the correct answer on the right.

_____ 1. Albert Einstein is one in science.

    a. number-crunching

_____ 2. Computers are very good at this.

    b. get bogged down

_____ 3. New Zealand is this if you live in Canada, but not if you live in Australia

    c. under way

_____ 4. Don't do this with any one question on a long exam!

    d. heavyweight

_____ 5. Spring has come, and training is _____ for the summer soccer competition.

    e. far-flung

II. Based on the text you just read, choose the best definition for the underlined word or phrase. The words and phrases are also underlined in the text for your reference. The form or part of speech of a word in this exercise may be different from the one used in the text. If you aren't able to determine the meaning through context, use other vocabulary skills such as word parts and word forms to help you with the meaning. <u>Note</u>: *The word* some *in 7 is followed by three correct meanings; choose the only one that is correct for the article you just read in this section.*

1. <u>to pioneer</u>:	a. study socialism	b. be the first to do	c. explore the wild
2. <u>to implement</u>:	a. carry out	b. repair	c. contact
3. <u>incompatible</u>:	a. not matching	b. illegible	c. incomprehensible
4. <u>to invoke</u>:	a. interpret	b. purchase	c. activate
5. <u>virtually</u>:	a. jokingly	b. very nearly	c. honestly
6. <u>unprecedented</u>:	a. not always	b. never before	c. unbelievable
7. <u>some</u>:	a. a little	b. a lot	c. approximately
8. <u>upsurge</u>:	a. sudden increase	b. giving up	c. risky operation
9. <u>instrumental</u>:	a. playing music	b. keeping quiet	c. playing an important role
10. <u>to advocate</u>:	a. advertise	b. recommend	c. criticize
11. <u>idle</u>:	a. not in use	b. very powerful	c. very small
12. <u>seamless</u>:	a. not obvious	b. undeveloped	c. perfectly smooth

What's the Point? Section 4

Demonstrate your understanding of some of the details of the reading. Based on the text, give short answers to the following questions.

1. When can a software bug be dangerous?

2. Is the process of software development highly disciplined today?

3. What are Nancy Lynch and Stephen Garland developing?

4. Where do Lynch and Garland work?

5. Is anyone in the computer industry currently working on software assurance?

6. What is meant by abstraction, in paragraph 3 of Section 4?

7. What do different software-assurance methods have in common?

8. What is unique about Lynch and Garland's method?

9. Do computer scientists believe they can achieve error-free software?

10. Why is it helpful to generate computer programs automatically?

💡 Understanding Words and Phrases: Section 4

I. Complete the following sentences in your own words, or provide answers to the questions. The vocabulary words are underlined; they are also underlined in the article for your reference.

1. Some <u>consequences</u> of a poor diet are _____.

2. To whom would you be willing to <u>reveal</u> your secrets?

_____

3. Her <u>rigorous</u> medical training included _____.

4. I cannot <u>reason</u> with him when _____.

5. What <u>corrective action</u> might be taken if a student misbehaved in school? _____

6. It is <u>critical</u> for a judge to be _____.

II. *The words in the left column can be associated with the items on the right; the items on the right are not definitions. Match the expressions on the left with their associations on the right. The expressions on the left are underlined in the article text for your reference.*

_____ 1. yield	a. a car accident
_____ 2. launch	b. a staircase
_____ 3. disaster	c. a new television program
_____ 4. collision	d. results
_____ 5. ascend	e. earthquake
_____ 6. descend	f. numbers from 10 to 1

What's the Point? Section 5

Check your understanding of the text. Based on the reading, are the following statements true (T) or false (F)?

1. _____ Quantum cryptography is more powerful than any existing encoding method.

2. _____ Scientists began to research quantum cryptography only a couple of years ago.

3. _____ A commercial quantum cryptography system is not available yet.

4. _____ In quantum cryptography, the key to a secret code consists of photons of light.

5. _____ With this technology, users will always notice when a code key has been stolen.

6. _____ Secret codes used today can be easily decoded by today's computers.

7. _____ A big problem with quantum cryptography is that encoded information travels over limited distances.

8. _____ Nicolas Gisin is developing quantum cryptography in the United States.

9. _____ Gisin's work will be useful only if fast computers of the future can break today's codes.

Understanding Words and Phrases: Section 5

Choose the appropriate completion for each of the following sentences. The underlined word in each sentence is also underlined in the article text for your reference.

      a. waiting an hour for her friend

      b. a constant smile

      c. fit through the front door

      d. heavy rain and wind

      e. vitamins and minerals

      f. fell asleep in class

      g. running the family business

      h. traveling around the world

1. When we moved into the house, the wide sofa scarcely _____.

2. Rice and flour are often fortified with _____.

3. Alice's patience began to wear thin after _____.

4. After staying up all night, Roger inevitably _____.

5. They trusted their son and left him in charge of _____.

6. A clown's makeup creates the illusion of _____.

7. The retired couple were excited at the prospect of _____.

8. A cave sheltered the mountain hikers from _____.

Now, write your own sentences using the underlined words in the sentences on page 175.

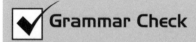 **Grammar Check**

Compound Adjectives

Every section of this article contains many examples of compound adjectives. A compound adjective consists of two or more words that are connected by a *hyphen*, a short dash. Here are several examples from the text.

lab-grown alternatives	**log-in** procedure
polymer-and-stem-cell mixture	**error-free** software
open-source implementation	**random-number** generator

The hyphenated words act as a single unit, a single adjective. Without a hyphen, each modifier that precedes a noun applies separately to the noun. For example, *big red house* means that the house is big and the house is also red. However, in the phrase *random-number generator,* the numbers are random, but the generator is not random; the generator creates random numbers.

Furthermore, some words that precede a noun may not modify that noun at all. Compare *polymer and stem-cell mixture* with *polymer-and-stem-cell mixture.* In the first case, there could be two materials: (1) a polymer, and (2) a mixture that contains only stem cells but no polymer. In the second case, there is only one material: a single mixture that contains both a polymer and stem cells. The hyphens are necessary to clarify the meaning. Without a hyphen (or hyphens), such a phrase would be ambiguous—it could have two different meanings.

In this exercise, several ambiguous noun phrases are listed. The last word in the phrase is the modified noun. Give two possible meanings of each phrase: (a) without a hyphen (or hyphens) between the modifiers and (b) with a hyphen (or hyphens). The first two noun phrases are done for you.

1. modern art history: a. modern art history = modern history of art
   b. modern-art history = history of modern art

2. fish and potato pie: a. two dishes: (1) fish, (2) potato pie
   b. one dish that contains both fish and potatoes

3. Italian language student: a. _____
   _____
   b. _____
   _____

4. large family car: a. _____
   _____
   b. _____
   _____

5. French bread baker: a. _____
   _____
   b. _____
   _____

6. beef and vegetable soup: a. _____
   _____
   b. _____
   _____

7. old furniture store: a. _____
                             _____

                          b. _____
                             _____

8. two family houses: a. _____
                             _____

                          b. _____
                             _____

 Let's Talk about It

1. How are these five emerging technologies different from gadgets and gizmos?

2. At the end of section 2, "Nano Solar Cells," what is meant by the phrase "from a boutique source to the Wal-Mart of electricity production"?

3. Each section of this article contains a table titled "Others in [the field]." Look at these tables. What types of organizations—private companies, universities, government—are doing most of the research? Can you explain this?

4. Do you agree that the technologies described in this article will be part of our future?

   a. Which technologies are likely to be developed and used? Why?

   b. Which ones are not likely to be successful? Why?

5. About 50 years ago, some people made predictions about the future that did not come true. For example, they predicted that today (a) most of our electricity would come from clean nuclear power, and (b) we would use cheap, disposable, biodegradable plastic dishes made of natural plant materials. These technologies actually exist today, but we don't use them as widely as predicted. Why is that?

   a. Where is nuclear power used? Why don't we use it more widely?

   b. Why do we still use traditional dishes?

6. Do you think that people's feelings—such as fear, beliefs, nostalgia, or personal taste—affect what new technologies we accept and use? How so?

7. Beside changes in technology, can you think of other changes that have happened over the last 50 years?

    a. What changes have we seen in our environment?

    b. How about our society?

8. Do you think that environmental and social changes can influence what technologies we develop and use? How?

What Do You Think?

1. What are your thoughts and feelings about technology and society? Choose any one of the discussion questions or issues on page 178 and write about your reactions to it. Explain your opinions and feelings, supporting them with facts and logical arguments.

2. Imagine life 50 years from now. What will people talk about? Write a dialogue that takes place in the year 2050. The conversation can be between parent and child, doctor and patient, two students, or anyone you choose. In this conversation, demonstrate the impact that a technological or social change has had on the lives of your characters.

3. It's 2050 and you are a news reporter. Write an article describing a major news story or several news events of the day. Persuade your readers that this news item is important. How could it affect individuals or society?

Expansion Activities

Create your own crossword puzzle. List all the new words and phrases that appeared in this unit. Make a crossword puzzle using as many of these words as you can. Feel free to add other words to fill in more of the puzzle space. Write your own clues for the words in your puzzle. The clues can be direct definitions, sentences with omitted words, or other descriptions. Give the puzzle to your classmates to solve.

List of Vocabulary Practiced

This list contains all the words and phrases that appear under *Key Words* and in vocabulary exercises in this book. The words and expressions are listed in alphabetical order, with the number of the unit and in some cases with the number of the section where they appear in parentheses.

abundant (7)

adapt/adaptable (5)

adhesive (5)

advocate (9, Section 3)

ailment (6)

align (9, Section 2)

alter/altered (1)

anticipation (6)

appear (1)

application (5)

arcane (9, Section 3)

ascend (9, Section 4)

assemble (8)

atrophy (3)

attempt (3)

autonomous (7)

bear (5)

benefit (1)

bind (8)

blade (4)

bond (5)

boost (9, Section 2)

bounds (9, Section 1)

branch (5)

break (a code) (9, Section 5)

break down (3)

breakdown (3)

breakthrough (4)

cadaver (9, Section 1)

capacity (4)

cartilage (9, Section 1)

cell (9, Section 1)

charge (9, Section 5)

chronological (6)

civil engineer (9, Section 4)

claim (5)

cling (5)

clump (5)

cluster (6)

collaborate (3)

colleague (5)

collision (9, Section 4)

component (2)

composite (9, Section 2)

concentrate (8)

concoct (9, Section 2)

confine/confinement (3)

consequence (9, Section 4)

considerable (8)

contribute (5)

conventional (4)

coordinate/coordination (6)

corrective action (9, Section 4)

counterpart (8)

countless (5)

crack (a code) (9, Section 5)

crash (9, Section 4)

critical (9, Section 4)

crop (9, Introduction)

crucial (4)

cryptography (9, Section 5)
dangle (5)
decryption (9, Section 5)
defect (6)
deflate (8)
deliberations (7)
delighted (7)
density (5)
departure (4)
descend (9, Section 4)
determine (3)
disaster (9, Section 4)
disperse/dispersal (9, Section 3)
display (1)
dissolve (8)
dwarf (9, Section 2)
dynamite (8)
electrode (9, Section 2)
emerge (9, Introduction)
emit (8)
employ (2)
emulate (5)
enable (5)
encompass (3)
encryption (9, Section 5)
enhance (9, Section 1)
enormous (7)
entrepreneur (9, Section 5)
envision (4)
envy (5)
evacuate/evacuation (9, Section 3)
exacting (9, Section 2)
exceed (2)
exploit 9 (Section 1)
extract (8)
faint (8)
feasibility (9, Section 1)
flexible (4)
float (2)

force (2)
fortify (9, Section 5)
fossil (8)
friction (2)
frontier (9, Section 1)
gadget (9, Introduction)
generate (2)
govern (2)
harmful/harmless (3)
harness (9, Section 2)
heavyweight (9, Section 3)
hinge(d) (4)
hormone (3)
hospitable (7)
idle (9, Section 3)
illusion (9, Section 5)
implant (9, Section 1)
implement (9, Section 3)
incompatible (9, Section 3)
incorporate (3)
indication (3)
inevitably (9, Section 5)
inflate (2)
infuse (9, Section 1)
inject (9, Section 1)
instant/instantly (1)
instrumental (9, Section 3)
integrate (6)
intermittent (4)
interval (6)
intricate (5)
invasive (9, Section 1)
invert (1)
invoke (9, Section 3)
key in/keyed in (1)
launch (9, Section 4)
mandate (8)
massive (8)
mimic (3)

mold (5)

monitor (8)

mount (5)

mutter (9, Section 4)

nestle (5)

neutralize (8)

niche (4)

number-crunching (9, Section 3)

obsolete (8)

overcome (2)

perceive/perception (6)

permanent/permanently (1)

phobia (5)

photovoltaic (9, Section 2)

pin down (5)

to pioneer (9, Section 3)

popular (1)

precede (1)

profound (7)

prone (4)

property (5)

prospect (9, Section 5)

protein (3)

protocol (9, Section 3)

prototype (4)

pseudo (9, Section 4)

pursuit (7)

rapid (1)

raw (3)

readily (5)

reason (9, Section 4)

reboot (9, Section 4)

repel (5)

retain (2)

retard (6)

reveal (9, Section 4)

revolutionary/revolution (1)

rigorous (9, Section 4)

sample (3)

scaffold (9, Section 1)

scamper (5)

scarcely (9, Section 5)

scurry (5)

seamless (9, Section 3)

sequence (1)

shelter (9, Section 5)

shift (4)

shortcoming (5)

skim (xiii)

software bug (9, Section 4)

solo (2)

some (9, Section 3)

sophisticated (1)

split (5)

stand for (1)

stiff (8)

straightforward (8)

struggle (7)

subject (3)

substance (8)

subtle (5)

suit/suited (9, Section 2)

supplement (3)

sweep/swept (2)

sweltering (7)

synchronization (4)

   in sync; out of sync (4)

synthesis (3)

synthesize (3)

syringe (9, Section 1)

tell-tale (8)

temporary (1)

thermal (8)

tissue (9, Section 1)

top-end (9, Section 2)

transform (8)